THE NERVOUS MECHANISM OF PLANTS

THE NERVOUS MECHANISM OF PLANTS

BY

SIR JAGADIS CHUNDER BOSE

GYAN

Published by

GYAN

Gyan Publishing House
5, Ansari Road
Daryaganj, New Delhi-110002
Phone: 011-47034999, 9811692060
E-mail: books@gyanbooks.com

Distribution Network
gyanbooks.com
India, USA, Canada, UK, Australia

ISBN: 978-93-6280-444-0 (HB)
First Published, 1926

2nd Impression 2024

Printed at: Gyan Press, Delhi.

THE NERVOUS MECHANISM OF PLANTS
Author: Sir Jagadis Chunder Bose

THE NERVOUS MECHANISM OF PLANTS

BY

SIR JAGADIS CHUNDER BOSE

M.A., D.SC., LL.D., F.R.S., C.S.I., C.I.E.
DIRECTOR, BOSE RESEARCH INSTITUTE
CALCUTTA

WITH 82 ILLUSTRATIONS

LONGMANS, GREEN AND CO. LTD.
39 PATERNOSTER ROW, LONDON, E.C. 4
NEW YORK, TORONTO
BOMBAY, CALCUTTA, AND MADRAS
1926

To

MY LIFE-LONG FRIEND

RABINDRA NATH TAGORE

PREFACE

My investigations on plant-response date from the discovery of the electric response of non-living matter, such as metals, to stimulus, published in 1900 by the International Congress of Science, Paris. The response, like that of living matter, was shown to exhibit fatigue under continuous stimulation, enhancement under chemical stimulants, and permanent abolition under poisons. These results indicated that the response of the more complex and unstable living matter is ultimately the expression of physico-chemical reactions. I next tried to find whether ordinary plants, meaning those usually regarded as insensitive, exhibit the characteristic electric response already known in 'sensitive' plants. Ordinary plants were regarded at the time as inexcitable, because they did not respond to stimulation by an obvious movement. In my Friday Evening Discourse before the Royal Institution in May 1901, I was, however, able to show that every plant, and even each organ of every plant, is excitable, and responds to stimulus by electric response of galvanometric negativity, the response being abolished at the death of the plant. A more detailed account of the results was published in my work on 'Response of the Living and Non-Living' (1902).

My next investigation was directed towards obtaining evidence of responsive mechanical movement in these plants rendered conspicuous by various devices of magnification ('Plant Response,' 1906). The effects of various environmental stimuli on different plant-organs were thus demonstrated by automatic records given by the plant. The most important fact established in plant-response was the nervous character of the impulse transmitted to a distance. My discovery of the excitatory polar action of an electric current and its transmission to a distance, proved that the conduction of excitation in the plant is fundamentally the same as that in the nerve of the animal. Further corroboration was found in the arrest of transmission by the application of electrotonic and other physiological blocks in the path of conduction.

In my 'Comparative Electro-Physiology' (1907) I employed the independent method of electric response and was able fully to confirm and extend the results which I had obtained by the method of mechanical response. Experiments are described showing that the response of isolated plant-nerve is indistinguishable from that of the animal nerve throughout a long series of parallel variations of condition. So complete, indeed, is the similarity, that the discovery of a responsive characteristic in a given case has proved a sure guide to its observation in the other. As an example of this may be mentioned the control of the nervous impulse in the animal by a homodromous or a heterodromous electric current, the discovery being due to the success which attended my attempt to control the nervous impulse in the plant by the directive action of electric currents. Accurate measurement of the normal velocity of transmission of nervous impulse, and its induced variations, has been rendered possible by my device of Resonant Recorder, described in 'Irritability of Plants' (1913), by which time-intervals as short as .005 second can be automatically recorded.

The great advance in animal physiology has been due to very sensitive and accurate methods in quantitative measurement, on which alone can any sound theory be based. In plant-physiology, unfortunately, no such methods had been previously available; this accounts for the unfounded speculations that had paralysed advance of knowledge in plant-physiology. As an instance of this may be mentioned the far-fetched theory recently advanced that the transmission of excitation in the sensitive plant, *Mimosa pudica*, is due to the movement of sap in the transpiration-current. Simultaneous measurements of the rate of transmission of excitation and of the ascent of sap show that the former is far greater than the latter. In Mimosa the velocity of nervous impulse in thin petioles is as high as 400 mm. per second, while the movement of sap is about 200 times slower. No demonstration of the unfounded character of the transpiration-current theory could be more simple and convincing than the observation of the effect of application of a drop of acid to the tip of the uppermost leaf of Mimosa described in Chapter II (p. 19).

My recent discovery of the transformation of the afferent or sensory into an efferent or motor impulse in the reflex arc in the pulvinus of Mimosa will, it is hoped, materially advance our knowledge of nervous impulse in general. I have, in the present work, given not only a connected account of my previous results which came out in scattered publications, but also a mass of new material that has been accumulated since the discovery of separate nerves for the conduction of sensory and motor impulses.

The results of the investigations which I have carried out for the last quarter of a century establish the generalisation that the physiological mechanism of the plant is identical with that of the animal. For there is hardly any phenomenon of irritability observed in the animal which is not also discoverable in the plant. In the multicellular animal organism as higher complexity was attained, it was accompanied by the gradual evolution of a nervous system, by which the different organs are put in intimate connection with each other and their various activities co-ordinated for ensuring the common good of the organism. Such connecting nervous links had not been suspected in the plant, commonly regarded as distinctly lower in the scale of evolution. The researches described in the present work show that not only has a nervous system been evolved in the plant, but that it has reached a very high degree of perfection, as marked by the reflex arc in which a sensory becomes transformed into a motor impulse. The characteristics of the two impulses, and the definitely distinct channels for their conduction, can be studied with greater certainty and accuracy in the plant than in the animal. And it may be confidently expected that the broader outlook of the unity of physiological mechanism in all life will lead to a great advance in the physiological investigation of the irritability of all living tissues.

The wide interest that has been roused in the new methods of investigation and their results, has been a matter of much gratification to me. I take this opportunity of tendering my best thanks to Prof. E. Pringsheim for preparing the German edition of the 'Physiology of the Ascent of Sap,' and also to Messrs. Gauthier-Villars for publishing a French edition of most of my works. My acknowledgments are also due to my research assistants and scholars for the very efficient help rendered by them.

J. C. Bose.

Bose Institute, Calcutta.
October 1925.

CONTENTS

CHAPTER I

INTRODUCTORY

CHAPTER II

THEORIES OF HYDRO-MECHANICAL TRANSMISSION AND OF TRANSMISSION BY THE TRANSPIRATION-CURRENT

CHAPTER III

EXCITATORY CHARACTER OF THE TRANSMITTED IMPULSE

PAGE

CHAPTER IV

ANATOMICAL CHARACTERISTICS OF THE CONDUCTING TISSUE

CHAPTER V

TRANSMISSION OF EXCITATION ACROSS A SYNAPTIC MEMBRANE

CHAPTER VI

AUTOMATIC RECORD OF VELOCITY OF TRANSMISSION IN PLANTS

CHAPTER VII

EFFECT OF EXTERNAL CHANGES ON THE VELOCITY OF TRANSMISSION

CHAPTER VIII

POSITIVE AND NEGATIVE IMPULSES UNDER INDIRECT STIMULATION

CHAPTER IX

ELECTRIC RESPONSE OF ORDINARY PLANTS TO DIRECT AND INDIRECT STIMULATION

CHAPTER X

ELECTRIC RESPONSE OF MIMOSA TO INDIRECT STIMULATION

CHAPTER XIV

THE DIA-HELIOTROPIC ATTITUDE OF LEAVES

CHAPTER XV

THE REFLEX ARC IN MIMOSA AND THE TRANSFORMATION OF AFFERENT INTO EFFERENT IMPULSE

CHAPTER XVI

SEPARATE CONDUCTING NERVES FOR SENSORY AND MOTOR IMPULSES, AND THE 'LOST TIME' IN REFLEX

CHAPTER XVII

ILLUSTRATIONS

THE NERVOUS MECHANISM OF PLANTS

CHAPTER I

INTRODUCTORY

IN the body of a multicellular organism necessity arises for intercommunication and interaction between the more or less distant organs. I have shown elsewhere [1] that this is accomplished in the plant, as in the animal, in two different ways—by translocation of matter, and by transmission of motion. The first is effected by the *slow* movement of fluids carrying chemical substances in solution, such as occurs in the circulation of sap: the second, by the *rapid* propagation of protoplasmic excitation, such as the nervous impulse in the animal. Great confusion has arisen in plant-physiology from want of discrimination between the two types of transmission, physical convection and physiological conduction.

Confining our attention first to the true excitatory impulse, it is conducted in the animal by a definite tissue known as nerve, the main function of which is the rapid transmission of excitation to a distance. When the animal nerve is pinched, or irritated in any way, an invisible molecular disturbance is propagated along it, and the transmitted impulse, impinging on the terminal motor organ,

[1] *Physiology of the Ascent of Sap* (1923), p. 269.

produces the well-known muscular twitch. The communicating system may, for convenience, be distinguished into three parts: the point where the stimulus impinges is the *Receptor*; the channel for the transmission of excitation is the *Conductor*; and the terminal motor organ which serves as an indicator is the *Effector*.

Stimulation is found to give rise in *Mimosa pudica* to an impulse very similar to the nervous impulse in the animal. Like the nervous impulse, it causes no visible change in the tissue which conducts it; it is conducted to a distance and, impinging on the motor organ, the pulvinus, causes the fall of the leaf. Other 'sensitive' plants present similar phenomena, among which may be mentioned *Neptunia oleracea*, *Biophytum sensitivum*, *Averrhoa Carambola*, and the arborescent *Mimosa Spegazzinii*. I have, in all my typical experiments, used *Mimosa pudica*, the other sensitive plants being occasionally employed for purposes of independent confirmation.

Modes of Stimulation

Mechanical stimulation.—This is effected by a superficial scratch. The Scratch-Stimulator (fig. 1, a) consists of a holder, from one end of which there projects an adjustable pin-point, the usual length of projection being only a fraction of a millimetre. A minimal stimulus is produced by a single scratch. The effective intensity of the stimulus can, however, be increased from minimal to maximal by the additive effect of repeated scratches following rapidly one after another. The scratches, it should be remembered, are purely superficial.

Electrical stimulation.—The two electrodes of an induction-coil are applied on the epidermis, electric connection being secured by means of kaolin-paste moistened with normal saline. The intensity of the stimulus can be increased from the minimal to the maximal by moving the secondary coil nearer to the primary.

Effect of Unilateral Stimulus of Moderate Intensity

A young and vertically straight specimen of Mimosa, growing in a pot, was used in the following experiments. In straight and untwisted specimens the stem bears the leaves in alternate series. Those of the odd series to the right are vertically over each other; so are also the leaves of the opposite even series to the left. In order to prevent shaking during the application of stimulus, the stem is securely held by a clamp without exerting undue pressure.

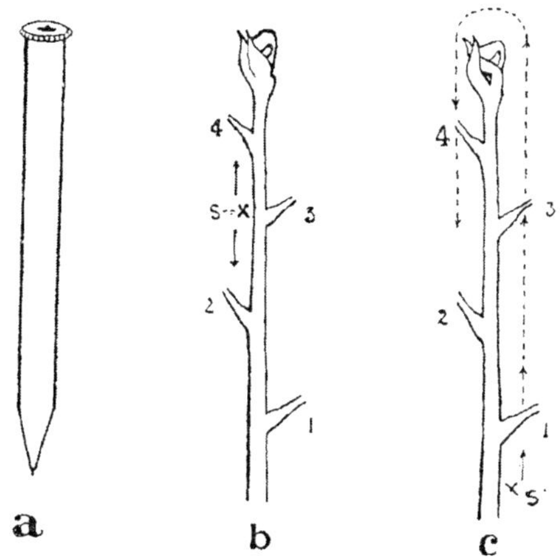

Fig. 1. Effect of Unilateral Scratch Stimulation.

a, The Scratch-Stimulator; b, effect of stimulation of moderate intensity s applied to left; c, effect of strong stimulus s′ applied to right side.

Experiment 1. *Mechanical Stimulation.*—Scratch-stimulus was applied to the stem on the left, between the second and the fourth leaves, at a point situated on the vertical line that passed through the two leaves. This gave rise to impulses which were conducted *simultaneously both upwards and downwards*, and caused the fall of leaves 2 and 4 (fig. 1, b). The distance between the point of stimulation S and the upper leaf 4 was 37 mm., the time of transmission being 10 seconds. The time of transmission downwards through 28 mm. was 20 seconds. It is to be noted that under unilateral stimulation, conduction occurred only on the stimulated side, the leaves on the opposite side remaining unaffected. When the stimulation was transferred from the left to the right side, the impulse was conducted only on the right side.

The velocity and the possible distance of transmission depend (1) on the physiological vigour of the plant, (2) on the season, (3) on the age of the internode, a young internode conducting more actively than an old, and (4) on the direction of propagation. It will be shown later (p. 48) that there is a preferential direction of conductivity along which the impulse travels at a quicker rate. The velocity of impulse in the stem varies from 2 to about 12 mm. per second.

Experiment 2. *Electric stimulation.*—Results in every way similar to those produced by mechanical stimulation were obtained with moderate electric stimulation. The impulse travelled both upwards and downwards, and the excitation of the leaves remained confined to the stimulated side.

Effect of Unilateral Stimulation of Stronger Intensity

A very remarkable effect is observed when the intensity of stimulus is increased from minimal to sub-maximal. This is attained in mechanical stimulation by the additive effect of repeated scratches, in electrical stimulation by the nearer approach of the secondary to the primary coil. I proceed to give a detailed account of a typical experiment on the effect of unilateral mechanical stimulation of moderately strong intensity. The impulse generated travelled both above and below the stimulated point. For the present purpose it is only necessary to follow the track of the ascending impulses.

Experiment 3.—A sub-maximal scratch-stimulus was applied unilaterally at S′, to the right, and vertically below leaf 1 (fig. 1, c). The ascending impulse first caused the fall of leaf 1, in the course of 20 seconds. The impulse then reached leaf 3, which fell 5 seconds after leaf 1. The impulse did not, however, stop at the highest leaf on the right side, but crossed over from the right side to the left, presumably after reaching the apex. The topmost leaf on the left side, numbered 4, fell 15 seconds after the fall of 3. The impulse which had hitherto been ascending on the right side, became converted after crossing at the apex into a descending impulse on the left side, as seen by the fall of leaf 2, 20 seconds after the fall of 4. Detailed results are given in the following tabular statement.

Intervening distance		Successive time-intervals
From S to 1 . .	15 mm.	20 seconds
,, 1 to 3 . .	60 mm.	5 ,,
,, 3 to 4 (via apex)	55 mm.	15 ,,
,, 4 to 2 . .	65 mm.	20 ,,

The results obtained with the stimulus of induction-shock are in every way similar to the above. It will be shown later (p. 22) that similar effects are also produced by the stimulus of polar action of a constant current. The results are therefore independent of the mode of stimulation, provided the stimulus is moderately strong. Excitation becomes outspread under excessively strong stimulus; I shall speak of this later.

I wish to draw special attention to the results of the fundamental experiments described above, the detailed consideration of which will help to a proper understanding of the true nature of the transmission of excitation. The important facts that have been established are given below.

(1) An impulse is generated by various modes of stimulation which are known to be effective in initiating nervous impulse in the animal.

(2) A superficial scratch, unattended by any exudation of sap, is effective in giving rise to an impulse. Irritation of the wood in the interior is not essential to the initiation of an impulse.

(3) Effective stimulation of the intact plant can be

effected by electric stimulus. There is, in this case, not the remotest possibility of exudation of sap nor of injury to the wood.

(4) Stimulation at any point in the vertical line between two leaves gives rise to impulses which travel simultaneously both upwards and downwards. The nervous impulse in the animal travels in both directions: the impulse in the plant does the same.

(5) The conducted impulse is confined to the stimulated side. It would thus appear that there are two definite conducting strands situated on opposite sides of the stem.

(6) Under stronger unilateral stimulation the ascending impulse on one side, after reaching the apex, crosses over to the opposite side and is converted into a descending impulse. This can only be explained on the supposition that there are two main conducting strands on opposite sides of the stem which meet at the apex.

The above inferences, which follow naturally from the sequence of excitatory fall of the leaves, will be found fully confirmed by the results of anatomical investigation given in a later chapter (p. 40).

Conduction between Stem and Leaves

That there is a continuous conducting channel between stem and leaf is shown by the successive fall of the leaves after stimulation of the stem. The converse of this, namely, conduction from leaf to stem, is well shown in the following experiment. A very simple and satisfactory means of *local* stimulation is found in the use of a hot glowing point. A flame is unsatisfactory for various reasons, one of which is that the rising hot air affects other leaves in the neighbourhood. The thin and long incense stick of the Chinese, when once lighted at one end, burns slowly without any flame, and requires no further care. The burning tip is always a glowing point and can therefore be applied for strong stimulation at any point on the plant. When a sub-petiole of Mimosa carrying the sensitive leaflets is thus locally stimulated, the course of the impulse is easily followed: (1) by the successive closure of the leaflets upwards; (2) by the drawing together of the sub-petioles due to excitation of the secondary pulvini; and (3) by the fall of the leaf due to excitation of the main pulvinus. The impulse then passes into the stem, causing the fall of other leaves. The existence of a definite conducting tissue which connects the leaf with the stem will become still more clear from the anatomical account given in a subsequent chapter.

With the very definite results that have been described it will be possible to remove the prevailing misconceptions in regard to the true nature of the transmission of impulse in the plant.

SUMMARY

Different modes of stimulating the plant have been described which cause no injury, no exudation of sap, and no irritation of the enclosed wood. An impulse is nevertheless generated which is transmitted to a distance.

Unilateral stimulation of moderate intensity causes an impulse which is propagated simultaneously up and down, along the same vertical line.

The course of the impulse is confined to the side stimulated, no excitation being produced on the opposite side. The conducting tissue is therefore not diffuse, but is definitely distributed in the stem.

Stronger unilateral stimulus generates an impulse which, after ascending to the apex, crosses over to the opposite side. The ascending impulse on one side then becomes converted into a descending impulse on the opposite side. There must therefore be two main conducting strands which meet at or near the apex.

Conducting continuity exists between the stem and the leaf.

CHAPTER II

THEORIES OF HYDRO-MECHANICAL TRANSMISSION AND OF TRANSMISSION BY THE TRANSPIRATION-CURRENT

It is now possible to discuss the nature of the transmission of impulse in plants, which is commonly considered to be very different from that of the nervous impulse in the animal. Transmission in the plant has been attributed (1) to hydro-mechanical disturbance caused by stimulation, and (2) to the translocation of a chemical stimulant by the movement of the sap.

Theory of Hydro-Mechanical Transmission

The above theory is largely based on the two well-known experiments of Pfeffer and Haberlandt. In the former of these, the effect of strong stimulation was found to travel over a chloroformed part of the petiole of Mimosa. Pfeffer assumed that the vital conductivity of this portion, if any, must have been abolished, since chloroform is known to abolish motile excitability. In the experiment of Haberlandt some of the intervening tissue was killed by scalding; in spite of this, the impulse was found to be transmitted across the scalded area.

From these two experiments it was inferred that the impulse which was transmitted could not have been of a true excitatory nature. It was held, on the contrary, that the strong stimulation had given rise to a variation, either increase or diminution, of hydrostatic pressure. This variation of pressure, it was assumed, had been hydro-mechanically transmitted, and on reaching the distant pulvinus had inflicted on it a blow which had proved as effective as if a mechanical stimulus had been applied directly. It was in fact held that transmission in Mimosa is purely mechanical and is not a propagation of protoplasmic excitation.

Though Pfeffer and Haberlandt advocate the theory of hydro-mechanical transmission, they disagree in regard to the channel for the transmission of impulse. According to Pfeffer, it is the vessels of the wood which conduct it; Haberlandt, on the other hand, insists that it is not the xylem-vessels, but certain tubular cells in the phloem that are the means of transmission. I italicise the important passages in the following quotations.

'Dutrochet was the first to show that stimuli are conducted through the vascular bundles of *Mimosa pudica*, and he also came to the correct conclusion that the transmission was due to pulsation of water. Pfeffer subsequently showed that the stimulus was able to travel over chloroformed parts of the stem, and Haberlandt found that dead regions of the stem and leaf retained their conductivity some time after they had been killed. We are therefore fully justified in ascribing the transmission of stimulus to the movements and changes of pressure of water in the vascular bundles, and when a cut is made in the stem, *a stimulus is only exercised when the knife penetrates the vascular bundles and allows the escape of a drop of water.*' [1]

Dutrochet is often quoted in support of the contention that the wood alone is concerned in the transmission of impulse in Mimosa; I take the following account of his experiment from Biedermann's 'Electro-Physiology.'

'(Dutrochet) showed that the cortex was not involved by paring away a ring of it, when the conductivity of the twig remained unaffected. So, too, on removing the pith. The wood alone was involved without exception, or *more correctly the vascular system* (*bast and vessels*).' [2]

[1] Pfeffer—*Physiology of Plants*—English translation (1906), p. 24.
[2] Biedermann—*Electro-physiology*—English translation, vol. ii. p. 12.

The account given above does not support the contention that the dead wood[1] is alone concerned in conduction, since the living phloem remained with the wood. Haberlandt, as already stated, came to the conclusion that certain tubular cells in the phloem functioned in the transmission of impulse. The following quotation explains his views on the mechanical transmission of impulse:

'The stimulus-transmitting elements of *Mimosa pudica* are not situated as Dutrochet, Sachs and Pfeffer, among others, have supposed in the woody cylinder or in the hadrome portions of the primary bundles, but on the contrary occur in the leptome strands, where they take the form of elongated tubular cells arranged in longitudinal series. . . . The effects of incision show that stimuli are actually propagated in this system of highly turgescent tubes, and that the mode of transmission is a hydro-dynamic one. If, namely, one or more of the tubular cells are laid open by an incision in the stem or petiole, then cell-sap instantly escapes in the form of a drop of transparent liquid; immediately afterwards the nearest pulvinus carries out the characteristic movement.'[2]

I will now briefly criticise the conclusions arrived at by Pfeffer and Haberlandt from their respective experiments on the effect of narcotisation and of scalding on the conduction of impulse. As regards the former, it is extremely doubtful whether the conducting tissue in the interior can be effectively narcotised by the external application of an anæsthetic. The task would almost be as difficult as narcotising a nerve-trunk lying between muscles, by the application of chloroform on the skin outside! In the case of the plant it is conceivable that, after a very long application, a small quantity of the narcotic might by ab-

[1] The xylem-tissue is not completely dead, since there are scattered living elements in it. Though these are not true conductors of excitation, yet the 'block' or resistance offered by them is overcome under intense stimulation. This abnormal diffusion is, however, a phenomenon quite different from the normal conduction of excitation along definite nervous channels.

[2] Haberlandt—*Physiological Plant Anatomy*—English translation, p. 643.

sorption get across to the internal conducting-tissue; but narcotisation in such circumstances could only be partial, and this partial block might fail to arrest the impulse due to the intense stimulation caused by a deep wound (see also p. 37).

In Haberlandt's experiment the conducting tissue was supposed to have been killed by scalding. If this had really been the case, then it may be supposed that under an exceptionally strong stimulus a hydrostatic disturbance had been transmitted through the dead tissue and caused stimulation of the distant leaf, as a mechanical blow *de novo*. Strong doubt may, however, be entertained as to whether the tissue had really been killed throughout. In my own experience I find it extremely difficult to be sure of killing the internal tissue of an organ by scalding the outside.

Complications of Wound-stimulation.

Both Pfeffer and Haberlandt used a deep cut or wound to effect stimulation. The complications introduced by the employment of such a drastic method will be understood from the results of Kühne's experiment on conduction of excitation in a frog's nerve.

'The delicate nerve which enters the middle of the sartorius by one side, divides within the muscle so that the single fibres that constitute the bifurcation branch many times dichotomously. When Kühne threw the broad upper end of the muscle into heat rigor by dipping it into warm oil, the half which remained normal twitched *on cutting the rigored portion with scissors, showing that excitable nerve-fibres could still be mechanically excited between the rigored and dead muscle-fibres*, and thus carry the excitation centripetally into branches which divide above the rigored portion of the muscle.'[1]

This experiment is clearly an instance of transmission of excitation through the heat-rigored animal tissue parallel to Haberlandt's experiment on transmission through the

[1] Biedermann—*Electro-physiology*, vol. ii. p. 57.

scalded tissue of the plant. In both these cases it is probable that the scalded tissues, though under heat-rigor, were not really killed. An induced block of conductivity is, after all, relative. There may well be an effective physiological block for normal intensities of stimulation, which would, however, fail under abnormal intensities of stimulus such as that of a burn or a cut. In Kühne's experiment the intense excitation of scissors-cut failed to be arrested, though the conductivity of the nerve had been depressed under heat-rigor. Similar considerations will explain how the intense excitation caused by a burn or a cut may be transmitted through the narcotised or scalded areas in Mimosa.

Diffuse Excitation under the Action of Intense Stimulus

The most fruitful source of error in the investigation of the conduction of excitation is the employment of such drastic modes of stimulation as deep wounds or burns. Though the normal transmission, under the action of moderate stimulus, is confined to the well-defined conducting tissue, the excitation becomes diffused under intense stimulation. This is demonstrated in Experiment 18, described in Chapter IV. The same thing also occurs in the nervous conduction in the animal, where, under abnormally powerful stimulus, the excitation becomes widely diffused. The necessity for discarding crude and drastic methods of stimulation in researches on conduction will now have become obvious. The object of inquiry is not to find out that a violent disturbance becomes widely diffused, but to determine the nature of the transmission of excitation under normal modes of stimulation. By employing stimulus of graduated intensity, it should be easy to determine the character of the impulse generated by observing the effects of various physiological blocks in modifying the power of conduction.

The Transpiration-Current Theory

The other recently proposed theory is that conduction in Mimosa is effected by the transpiration-current carrying along the xylem-vessels some stimulating substance excreted by the wood in consequence of stimulation. Ricca [1] describes the following experiment in support of this theory. He cut across a shoot of *Mimosa Spegazzinii* and connected the two pieces by a tube filled with water. Intense stimulation of the lower piece of the stem by a flame is, according to him, often followed by the fall of the leaves in the upper piece. Application of an extract obtained from the tissue of the internodes to the cut end of the stem was also observed to cause fall of the leaves. Snow [2] obtained similar results with *Mimosa pudica.*

In order to establish the transpiration-current theory, it is necessary to prove, (1) that stimulation is effected only by irritation of the wood; (2) that a hypothetical stimulant is excreted as a consequence of stimulation; (3) that transmission can take place across a water-gap; and (4) that the stimulant is transported by the transpiration-current with the same speed as the impulse which causes successive fall of the leaves.

The Wounding of the Wood

Pfeffer and Haberlandt contend that 'a stimulus is only exercised when the knife penetrates the vascular bundles and allows the escape of a drop of water.' The escape of water is thus supposed to give rise to a sudden variation of pressure and to a consequent hydro-mechanical impulse. The transpiration-current theory assumes the irritation of the wood to be the antecedent of the excretion

[1] Ricca, U.—Soluzione d'un problema di fisiologia—*Nuovo Giorn. Bot. It.*, vol. 23 (1916). Solution d'un problème de physiologie—*Arch. Ital. d. Biol.*, vol. 65 (1916).

[2] Snow, R.—Conduction of Excitation in Stem and Leaf of *Mimosa pudica*—*Royal Soc. Proc., B.*, vol. 96, p. 349 (1924).

of a stimulant which, on translocation by the movement of sap, is the cause of fall of the leaves. Now the hypothetical structure on which the two theories are based inevitably falls to the ground if it can be shown that stimulation can be effected (1) without wounding the wood, and (2) without exudation of sap from the cut stem.

It is remarkable that the upholders of these two theories should have been blind to the obvious fact that stimulation can be effected without wounding the plant by a superficial scratch, by rough friction, or by an electric shock of only moderate intensity. I have shown in Experiment 1 that the impulse generated by a superficial scratch, which could not irritate the wood, was transmitted to a distance; there was no wound nor was there any escape of sap to cause a hydro-mechanical disturbance.

The following experiment shows that mere friction is sufficient to initiate the impulse.

Experiment 4.—The petiole of Mimosa is held within clamping jaws to prevent any mechanical shaking. A suitable contrivance permits quick unclamping for observation of the responsive fall of the leaf. The petiole is subjected to the stimulus of friction by rubbing it with emery paper. If the clamp be now quickly opened the leaf will be at first found to retain its normal position, and it is only after a short interval that it exhibits the excitatory fall. The delay is due to the time taken by the impulse to travel through the intervening distance.

Even more convincing is the result already described of the effect of electric stimulation of moderate intensity on the intact plant (Experiment 2). As already pointed out, there is in such a case not the remotest possibility of exudation of sap or of injury to the wood.

Both the theories of hydro-mechanical disturbance and transpiration-current are based on the supposition that a deep wound, causing an escape of sap or irritation to the wood, is essential to the transmission of impulse. I have just shown that transmission takes place, even in the complete absence of any wound, under stimulation by superficial friction or by electric shock. The essential structure on which the two theories are erected thus falls to the ground.

The theory of the transpiration-current involves three other assumptions: that transmission takes place across a water-gap; that the transpiration-current carries a hypothetical stimulant excreted by the stimulated wood; and that the velocity of the sap-movement is the same as that of the transmission of impulse. The following experiments will be found to disprove all three assumptions.

Transmission across a Water-gap

I may begin by mentioning that Koketsu[1] cut across the petiole of Mimosa and connected the two portions by a water-tight tube filled with water; no evidence was ever obtained that impulse generated by stimulation of the distal half was conducted across the water-gap to the other.

My own experiments with the stem of Mimosa to test the supposed transmission across a water-gap were carried out with more than fifty specimens of *Mimosa pudica*, grown in vigorous condition in the grounds of my Institute. The experiments included diverse modes of intense stimulation; yet the results in all cases were in perfect agreement. They were in some cases made with cut stems; I have shown elsewhere[2] that the normal excitability of the cut specimen becomes fully restored in the course of a few hours, and that the response of the cut specimen is then very similar to that of the intact plant.

Experiment 5.—A divided stem had its two cut ends enclosed in a water-tube, and the two ends were brought in contact with each other. The sensitiveness of the upper and lower portions of the stem was such that separate

[1] Koketsu, R.—*Journal of the Department of Agriculture*, Kyashu Imperial University, vol. 1, No. 1, p. 55 (1923): also Bose—Physiological and Anatomical Investigation on *Mimosa pudica*—*Proc. Roy. Soc.*, B. 98, 1925, p. 290.

[2] Physiological Investigations with Petiole-Pulvinus Preparations of *Mimosa pudica*—*Proc. Roy. Soc.*, vol. 89, 1916.

application of induction-shock of minimal intensity, lasting for one second, caused the fall of the leaves of both portions. After recovery, stimulus was applied to the lower piece, its intensity being increased to the maximal; the duration of application was also increased from one to twenty seconds. In spite of this very intense stimulation there was no evidence of conduction across the water-gap.

Experiment 6.—Stimulation was electro-thermal. A thin piece of platinum wire was wound round the lower piece of stem; the passage of a strong electric current made the wire red-hot. Though the lower piece was scorched by this intense stimulation, the upper piece remained unexcited, showing that no impulse was transmitted across the water-gap. This and the previous experiments were repeated with more than fifty different specimens; in not a single instance was there any evidence of transmission across the gap.

The supposed transmission across a water-gap, in the case of the stems and petioles of *Mimosa Spegazzinii*, led Ricca to formulate the transpiration-current theory. Snow, in his recent paper,[1] states that 'in the leaf of *M. Spegazzinii* excitation is regularly conducted by some mechanism that has nothing to do with the water-current.' In opposition to the very far-fetched theory of pressure-variation put forward by Dixon, Snow asserts that 'changes of pressure in the tube-cells play no part in conducting excitation, even in the leaf.' Snow is nevertheless of opinion that transmission in the stem is normally brought about by movement of water which carries the hypothetical stimulant. It will be presently shown that this supposition is without any foundation.

The Hypothetical Stimulant

The transpiration-current theory involves the assumption of the excretion of a stimulant substance as the result

[1] *Proceedings of the Royal Society*, April 2, 1925.

of stimulation. I have failed, as the following experiment shows, to obtain evidence that any such substance is excreted.

Experiment 7. *Application of extract at cut end.*—An extract prepared from internodes of Mimosa was applied at cut end of the stem. This caused no stimulation, though the experiment was repeated with twenty different specimens. It is conceivable that certain vegetable extracts might act as stimulants; vegetable alkaloids of a poisonous nature, for example, produce excitation when applied in minute doses to the cut end of the stem. It would, however, be quite unreasonable to conclude from this result that an alkaloid is excreted by the plant under the action of a minimally effective stimulus.

But even assuming the presence of a stimulating substance, the question of the velocity of its translocation still remains; for, if the transpiration-current theory be true, evidently the velocity of the transpiration-current which conveys the stimulant must be identical with that of the transmission of the impulse. Snow finds that the highest rate of ascent of sap in Mimosa is 18·5 cm. per minute, or 3 mm. per second. I find that the highest velocity of transmission of excitation in thin petioles of Mimosa is of a very different order, being as high as 400 mm. per second. Again, under normal conditions, the ascent of sap is upwards, but excitation travels both upwards and downwards simultaneously.

Instead of comparing the velocity of the movement of sap and of excitation in different specimens, the results would be more conclusive if their respective rates were determined in an identical specimen. It is moreover possible to arrange the experiment in such a manner that there is practically no movement of sap in the direction of transmission of excitation.

Simultaneous Determination of Movement of Sap and Transmission of Impulse

The rate of translocation of a stimulant can easily be ascertained (1) by the use of a staining agent which is also a stimulant; or (2) by the use of a colourless stimulating fluid, the extent of translocation of which by the movement of sap can be detected by the employment of a suitable 'developer.' I describe three typical experiments carried out with three different species of plants.

Experiment 8.—Methylene blue, in a moderately strong solution, is a stimulant. The thin and long flower-stalk of *Biophytum sensitivum* is a very efficient conductor of excitation. When the stalk is stimulated in any way by a mechanical or electrical stimulus or by a chemical irritant, the resulting impulse travels downwards, against the normal direction of ascent of sap, and reaches the rosette of leaves, causing the closure of the sensitive leaflets from the centre outwards. To hasten absorption of the stimulating stain, the top of the flower-stalk was cut across and a piece of moist cloth placed on the cut end. After recovery of the leaflets from the excitatory impulse caused by the cut, Methylene blue was applied at the cut end. The impulse initiated by the stimulant was found to reach the sensitive leaflets and cause their closure. The time required for transmission was 30 seconds, the intervening length of the flower-stalk and stem being 60 mm. The extent of translocation of the staining solution downwards was ascertained by examining the longitudinal section of the flower-stalk under the microscope. The translocation in 30 seconds was found to be quite negligible, certainly not more than through 0·3 mm. This was due rather to slow diffusion than to the movement of sap. Even on the assumption that the transpiration-current can carry the sap downwards, the rate of transmission of excitation was of a very different order to that of sap-movement, being at least 200 times quicker. I obtained similar results with Eosin stain.

Experiment 9.—The specimen in this case was *Mimosa pudica,* the stimulant employed being a drop of dilute hydrochloric acid. The extent of translocation was tested by application of silver nitrate solution, which produces a white precipitate. The acid was applied at the tip of the uppermost leaf. To promote absorption of the solution it is advisable to wash the tip with dilute ether. Half a dozen experiments gave very similar results, of which the following is typical. The impulse generated by the acid stimulant travelled downwards and caused the successive fall of seven lower leaves in the course of 40 seconds, the distance of transmission being 120 mm. Examination of the section of the plant with silver nitrate solution showed that the acid had not been translocated downwards, but remained practically localised at the point of application.

Experiment 10.—For this experiment I took a different species, the sensitive *M. Spegazzinii* of tree-like habit. The potted plant was young; the petiole of each leaf bore from five to seven pairs of sub-petioles, each of which in their turn bore about thirty pairs of relatively large-sized leaflets. The passage of impulse is easily followed by the excitatory fall of the sub-petioles and leaflets in regular sequence. The pulvinules of the leaflets are highly sensitive; the secondary pulvinus of the sub-petiole is also fairly sensitive; but the main pulvinus of the leaf is practically insensitive.

A drop of hydrochloric acid was applied to the extreme end of the uppermost sub-petiole to the right. The impulse travelled inwards and reached the main petiole; afterwards it travelled in a centripetal direction towards the stem, the successive pairs of sub-petioles and their leaflets exhibiting excitatory fall (fig. 2). The length through which the impulse travelled in the uppermost sub-petiole was 64 mm., the time of transmission being 10 seconds. The impulse reached four out of the five pairs of sub-petioles and was transmitted through a total distance of 135 mm. in the course of 55 seconds. Chemical examination with silver nitrate solution showed that the acid applied at the tip of the sub-petiole had not been translocated, but had remained localised at the point of application.

The results of the experiments described above prove conclusively that the transmission of impulse is in no way connected with the movement of sap. The theory of the

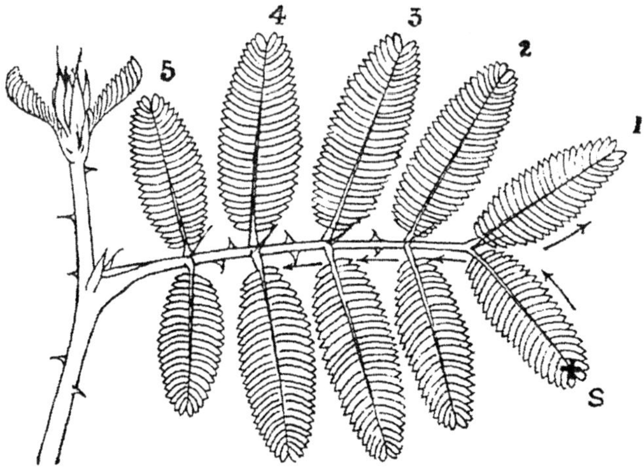

FIG. 2. Centripetal Impulse in Petiole of *M. Spegazzinii*, under application of chemical stimulus at S.

transpiration-current as the means of the conduction of excitation is therefore completely discredited.

Normal transmission is, then, not hydro-mechanical, nor is it due to movement of the sap. There thus remains but one alternative, that it is a propagation of protoplasmic excitation as in the nerve of the animal.

My investigations on the subject,[1] carried out for more than twenty years, afford conclusive proof of the nervous character of the transmission of impulse in plants. I propose to give in the following chapters a connected account of my

[1] *Plant Response* (1906); *Comparative Electro-physiology* (1907); *Irritability of Plants* (1913).

researches on the subject. I will first describe experiments in proof of the excitatory character of the transmitted impulse in Mimosa, and then deal with the characteristics of the conducting tissue. Finally I shall give an account of my recent discovery of a class of phenomena in Mimosa, which indicate a high degree of differentiation in its nervous system.

Summary

It has been contended, on insufficient grounds, that the wood is the tissue which conducts impulse. Haberlandt's experiments prove, on the other hand, that it is the phloem, and not the xylem, which is the conductor.

The hydro-mechanical theory is based on the observation of Pfeffer that stimulation is only effected when the knife has penetrated the vascular bundles and allowed the escape of a drop of water. The excitatory fall of the leaf at a distance is therefore supposed to be due to a transmitted mechanical blow or traction.

It is here shown that stimulation can be readily effected without a wound and escape of sap. The transmission is therefore not hydro-mechanical.

The theory of transmission by the transpiration-current is based on the supposition that a hypothetical stimulant, excreted as the result of irritation of the wood, is translocated by the transpiration-current with the same speed as the impulse which causes the successive fall of the leaves. Experiments have been made for the simultaneous determination of the velocity of the impulse and the rate of transport of the chemical stimulant which initiates it. The results show that while the chemical stimulant remained practically localised at the point of application, the impulse generated was transmitted to a considerable distance. The transmission of impulse is therefore in no way connected with the movement of sap.

CHAPTER III

EXCITATORY CHARACTER OF THE TRANSMITTED IMPULSE

THE results of the tests described in the last chapter show that transmission of impulse in Mimosa is due neither to a hydro-mechanical disturbance nor to the transpiration-current. I now offer evidence that transmission in the plant is physiological and is fundamentally of the same character as that of the nervous impulse of the animal.

Satisfactory evidence is obtainable only by the employment of a mode of stimulation which is purely physiological, and which causes no physical disturbance. Transmission of impulse must then be due solely to the propagation of protoplasmic excitation. Further confirmation would be afforded by effecting a physiological block which arrests an excitatory impulse but has no such effect on a hydro-mechanical disturbance or on the transpiration-current. The following are the results of my investigations on (1) the excitation induced by the discriminative polar action of a constant current, and (2) on the electrotonic arrest of transmission of impulse.

POLAR EXCITATORY EFFECT OF CONSTANT CURRENT

It is well known that an electrical current causes specific excitatory reactions in animal tissues. Excitation is characteristically produced at the sudden starting or 'make' of the current; it is also produced at the stoppage or 'break' of the current. Thus when a feeble electric current is passed by means of two electrodes through a muscle, an excitatory contraction is initiated on the sudden starting of the current at the kathode where the current leaves the tissue, no excitation being produced at the anode, that is at the point of entrance of the current. At the break of the current, excitation is produced neither at the kathode nor at the anode. When the intensity of the electric current is increased, the excitation due to the make still takes place only at the kathode; whereas at the break of the current excitation now takes place at the anode.

These characteristic effects are produced not only by the direct application of the current, but also by its indirect application. Thus a nervous impulse, initiated at the kathodic point of a nerve by the make of the current, is conducted along the nerve, causing excitatory contraction of the terminal muscle. Similarly the break of a stronger current initiates an impulse at the anode.

The characteristic polar effects of current on Protozoa were found by different observers to be, generally speaking, opposite to those in animal tissues. Hence it has been supposed that the laws of polar reaction in unfibrillated protoplasm must be different from those in highly differentiated animal tissues. My experiments on the polar action of electrical current on plants prove, on the other hand, that the reactions of the undifferentiated protoplasm of the plant-body are identical with those of the animal tissues. This is demonstrated by the following experiments with the primary pulvinus of *Mimosa pudica*.

POLAR ACTION OF CURRENT ON PULVINUS OF MIMOSA

Monopolar method.—One electrode from a battery is applied on the pulvinus and the second electrode attached to a distant indifferent point. Four cycles of operation can be performed with the commutator by which the current in either direction can be made or broken. Figure 3 illustrates the condition after sudden make of the current by tilting the commutator to the left, the pulvinus being made the kathode. By proper manipulation of the commutator, the pulvinus may be subjected (1) to kathode-make, (2) to kathode-break, (3) to anode-make, and (4) to anode-break.

Experiment 11. *Effect of feeble current.*—I used a leaf of Mimosa which was in a moderately sensitive condition. The electrical resistance between the two points of contact was found to be a million ohms. An E.M.F. of 8 volts was

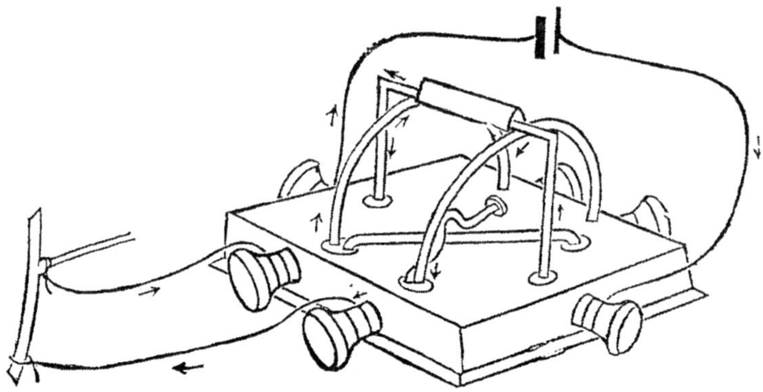

FIG. 3. The Commutator.

Electric connections are made with the pulvinus and a second indifferent point on the stem. Tilting the commutator to left makes the pulvinus kathode. Slight movement to right causes break of kathodic current. Tilting to the right produces make-anodic current at pulvinus. Slight tilting back produces anode-break.

found to be effective in inducing the excitatory fall of the leaf, the pulvinus being made the kathode. Excitation is induced by a sudden variation of the current and not during the continuance of the current. Hence, if after the excitatory fall, the current is continued, the leaf re-erects itself. The current is now broken; this induces no excitation. The commutator is next tilted to the right, the pulvinus being made the anode; this again induces no excitation, nor is there any excitation on break at the anode.

Experiment 12. *Effect of moderate current.*—The E.M.F. of the acting current was next increased to 12 volts. In carrying the experiment through the usual cycle of operations, it was found that the kathode excited at make and not at break. The effect of the anode, on the other hand, was different; it excited at break and not at make.

The automatic record by the plant is given in Fig. 4. The normal signal-line below indicates no current; the up-line, kathode-make; return to horizontal, kathode-break; down-line, anode-make; and return to the horizontal represents anode-break. It will be noted that under 8 volts, excitation takes place only at kathode-make. Under 12 volts there is a stronger response at kathode-make and none at kathode-break; no excitation was produced at anode-make, but it occurred at anode-break.

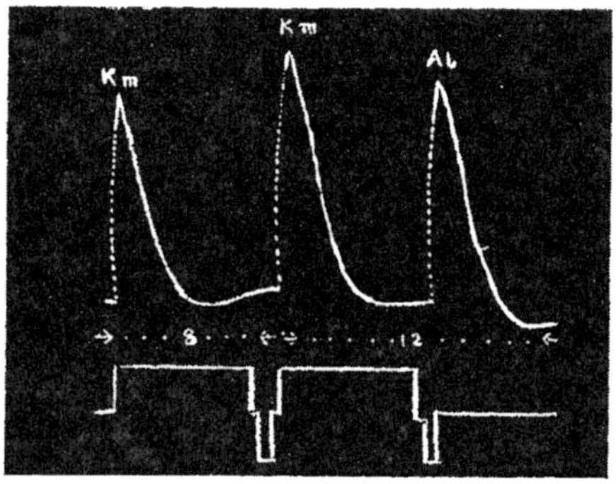

FIG. 4. Under 8 volts excitation at kathode-make Km; under 12 volts excitation at kathode-make and anode-break Ab.

The universality of these characteristic polar excitations is proved by the results of experiments which I carried out with the sensitive organs of various other plants, such as the leaflets of Mimosa, of Biophytum, of *Neptunia oleracea*, of *Averrhoa Carambola* and of *Averrhoa Bilimbi*. The effective intensity of the current depends on the excitability of the particular specimen, a vigorous specimen being excited by a feebler current. It also depends on the relative excitability of different motor organs, the pulvinules of the leaflets being, in general, more sensitive than the main pulvinus. As an example of the response of the leaflets, I describe an experiment with *Biophytum sensitivum*, employing the *Bipolar method*. The two electrodes were applied at two points on the petiole which carries numerous pairs of motile leaflets.

Experiment 13. *Effect of feeble current.*—This caused excitation only at kathode-make and not at break. No excitation was produced at anode on either make or break. The formula for excitatory reaction to feeble current is thus Km.

Experiment 14. *Action of moderate current.*—Excitation took place only at kathode-make and at anode-break. I give a sketch showing these effects (fig. 5); the figure to

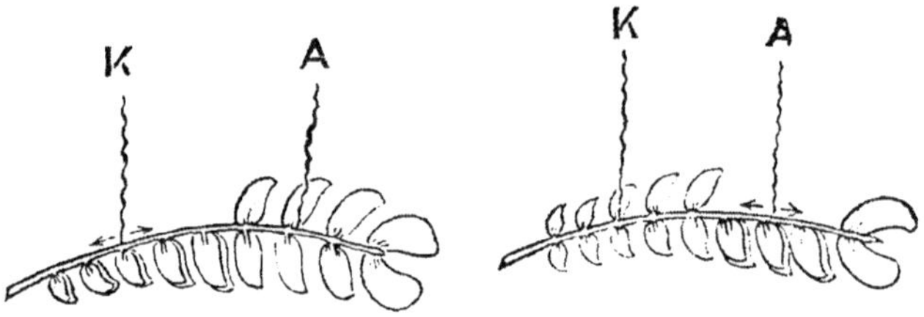

FIG. 5. Illustration to the left shows excitation induced by kathode-make, that to the right the effect of anode-break (*Biophytum sensitivum*).

the left shows the excitation initiated at the kathodic point at make, *which did not remain localised but was conducted in both directions.* After the recovery of the leaflets, the circuit was broken, and excitation was initiated at the point of anode-break. The formula for moderate intensity of current is KmAb.

Numerous experiments were carried out with different plants which gave similar results. In the two tables on p. 27 are given the effects of feeble and of moderate current. The value of the current given is what was found to be minimally effective for highly excitable specimens.

The specific reactions of plants to electric current are thus in every way similar to those of animal tissues.

The Laws of Polar Excitation for Plants are :—

(1) WITH FEEBLE CURRENT, THE KATHODE EXCITES AT MAKE AND NOT AT BREAK. THE ANODE EXCITES AT NEITHER MAKE NOR BREAK.

(2) WITH CURRENT OF MODERATE INTENSITY, THE KATHODE EXCITES AT MAKE AND NOT AT BREAK. THE ANODE EXCITES AT BREAK AND NOT AT MAKE.

TABLE I.—EFFECT OF FEEBLE CURRENT ON VARIOUS SENSITIVE PLANTS

Specimen	Number of experiments	Minimum current in micro-amperes
Leaflets of Mimosa .	25	0·7
,, Biophytum .	50	0·5
,, Neptunia .	5	7·0
,, *A. Carambola* .	10	4·0
,, *A. Bilimbi* .	5	6·0
Leaf of Mimosa .	15	2·0
Excitation occurs only on make at kathode. Formula Km.		

TABLE II.—EFFECT OF MODERATE CURRENT ON VARIOUS SENSITIVE PLANTS

Specimen	Number of experiments	Minimum current in micro-amperes
Leaflets of Mimosa .	25	1·6
,, Biophytum .	50	2·1
,, Neptunia .	5	12·0
,, *A. Carambola* .	10	11·0
,, *A. Bilimbi* .	5	16·0
Leaf of Mimosa .	15	4·0
Excitation only at kathode-make and at anode-break. Formula KmAb.		

The discriminative excitatory transmission on kathode-make and on anode-break proves that it is due neither to hydro-mechanical disturbance nor to the transpiration-current, but is a propagation of protoplasmic excitation.

The intensity of current initiating excitatory impulse is often extremely feeble; in Biophytum it may be as little as 0·5 micro-ampere, an intensity of current which cannot be detected even by the highly sensitive tip of the human tongue.

TRANSMISSION OF POLAR EXCITATION IN THE STEM

Experiment 15.—Two electric connections were made on the left side of an erect stem of *Mimosa pudica*,

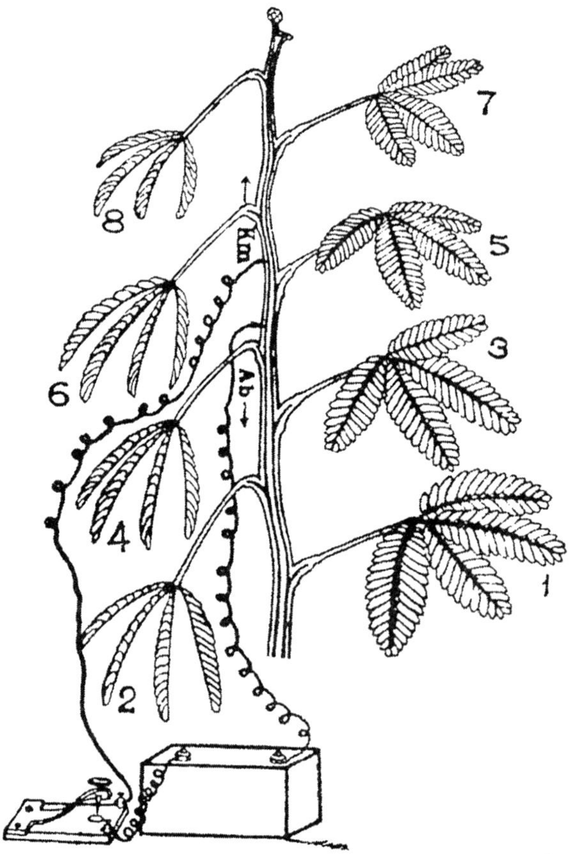

FIG. 6. Excitation at kathode-make Km and at anode-break Ab. The former travels upwards and the latter downwards.

at points intermediate between the leaves 4 and 6 (fig. 6). A fairly strong current of 6 volts was maintained. On starting the current an excitatory impulse was initiated at the kathode which was above, and leaves 6 and 8 fell in serial succession. Owing to electrotonic block, to be presently explained, the excitation could not be transmitted downwards. When the current was broken, excitation was induced at the anode below, and the impulse was transmitted downwards, the leaves 4 and 2 falling one after another.

Experiment 16. *Effect of a stronger current.*—The E.M.F. was increased to 10 volts. This gave rise to a stronger impulse at kathode-make, which ascended on the left side and caused the leaves to fall in sequence from below upwards. After the impulse had reached the apex, it crossed over to the right side, the direction of propagation becoming reversed from an ascending on the left side to a descending impulse on the right. The effect of moderately strong unilateral stimulation by the polar action of current is thus the same as those by scratch-stimulus, and by induction-shock (Experiment 3).

Arrest of Transmission by Electrotonic Block

Another method is available for determining the true nature of the transmitted impulse in plants. When a polarising current from a battery is maintained through a length of animal nerve, a transmitted nervous impulse is arrested; the physiological block persists during the passage of the current, conduction being restored on the cessation of the blocking current. It is of special interest to have thus at our disposal a physiological block which can be put 'on' or 'off' in succession. It is obvious that such a block could arrest neither the movement of sap nor a hydro-mechanical disturbance caused by a sudden variation of pressure.

Experiment 17.—In this investigation it is desirable to employ a testing stimulus which can be either maintained constant or increased in a quantitative manner. With the help of a sliding induction-coil it is easy to obtain an intensity of stimulus which is always effective. The proximal of the two exciting electrodes was placed on the main petiole, at a distance of 30 mm. from the primary pulvinus. Half way between the point of excitation and the pulvinus were placed the two polarising electrodes 5 mm. apart, through which a constant current could be maintained, producing the electrotonic block.

The record of the response of the leaf was obtained by means of the Resonant Recorder described in Chapter VI. The effective testing stimulus was always the same; it was applied at intervals of twenty minutes, by which time the leaf had re-erected itself. The electrotonic block was put 'off' and 'on' alternately. The signal below indicates the application of the test-stimulus; the block was put on at B and B. The record (fig. 7) shows that the transmitted excitation was invariably arrested whenever the blocking current was applied, and was restored on the cessation of the current.

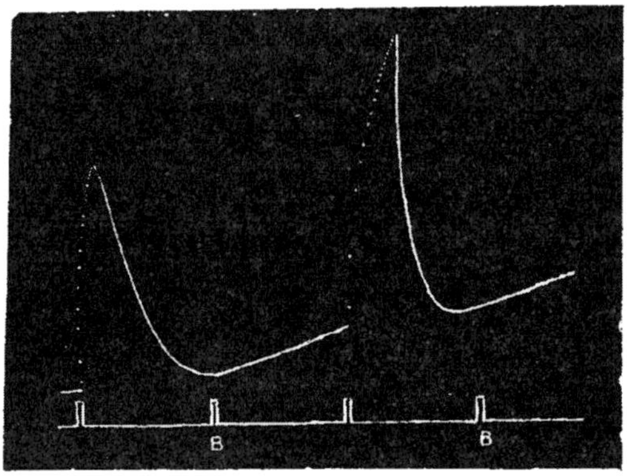

FIG. 7. Records of transmitted excitation with the block off and on. Arrest of transmitted excitation under electrotonic block B, B.

Inasmuch as the transmission of impulse in the petiole of Mimosa is arrested by an electrotonic block, as is the transmission in an animal nerve, the obvious conclusion is that transmission is essentially the same process in both; if it be called 'nervous' in the case of the animal, there is equal reason for applying to it the same term in the case of the plant. The excitatory impulse in plants will therefore be designated henceforth as the nervous impulse.

SUMMARY

Protoplasmic excitation is induced in plants by the polar action of a constant current. The impulse generated is purely physiological.

The following Laws of Polar Excitation in Plants have been established:

(1) WITH FEEBLE CURRENT, EXCITATION TAKES PLACE ONLY AT KATHODE-MAKE.

(2) WITH A CURRENT OF MODERATE INTENSITY, EXCITATION OCCURS AT KATHODE-MAKE AND AT ANODE-BREAK.

The polar reactions of the undifferentiated protoplasm of the plant-body are thus identical with those of highly differentiated animal tissues.

Excitation in certain sensitive plants is found to occur under a current so feeble that it cannot even be detected by the very sensitive tip of the human tongue.

The normal impulse in the conducting tissue of plants is therefore due not to any hydro-mechanical disturbance but to the propagation of protoplasmic excitation as in the nerve of animals.

Weak unilateral polar excitation is conducted only on the stimulated side of the stem; under stimulation with a stronger current, the impulse ascends on the stimulated side, and, after crossing at the apex, descends on the opposite side. The effect produced is similar to that under unilateral scratch-stimulation and stimulation by induction-shock.

The nervous impulse in an animal is arrested by the interposition of an electrotonic block, the conduction being restored on the cessation of the electrotonic current. Similarly, transmission of impulse in the plant can be repeatedly arrested and restored by the alternate application and removal of the electrotonic block.

These results conclusively prove that the transmission of excitatory impulse in plants is essentially similar to that of the nervous impulse in animals.

CHAPTER IV

ANATOMICAL CHARACTERISTICS OF THE CONDUCTING TISSUE

THE nervous impulse is itself invisible, and its propagation along the nerve can only be detected by the electric change which accompanies it. The nervous tissue is imbedded in the interior of the plant, and the problem of its definite localisation is surrounded by numerous difficulties. I have, however, been successful in solving it by my Electric Probe. When this is gradually introduced transversely into, say, the petiole, a sensitive galvanometer in circuit with the Probe remains practically quiescent until the tip of the Probe comes in contact with the nerve through which protoplasmic excitation is being transmitted; the passage of the impulse is then detected by an electric response of galvanometric negativity.

CONDUCTING STRANDS IN THE PETIOLE

The petiole contains four main vascular bundles shown in the micro-photograph of the transverse section (fig. 8): E is the epidermis; C is the cortical tissue; S is the hollow cylinder of sclerenchyma for ensuring mechanical strength and protection of the conducting tissue in the interior; F is one of the four vascular bundles; O is the central pith. The four vascular bundles converge at the pulvinus, and appear to coalesce into an almost continuous ring.

The following is a tabular statement of micrometric measurements of the distance of the different tissues from the surface and of their thickness.

TABLE III.—SHOWING DISTANCE FROM SURFACE AND THICKNESS OF THE DIFFERENT TISSUES IN THE PETIOLE

Tissue	Distance from surface	Thickness
Epidermis . . .	0 mm.	·02 mm.
Cortex . . .	·02 mm.	·09 mm.
Sclerenchyma . .	·105 mm.	·06 mm.
External phloem . .	·165 mm.	·12 mm.
Xylem . . .	·285 mm.	·13 mm.
Internal phloem . .	·415 mm.	·08 mm.
Pith	·495 mm.	

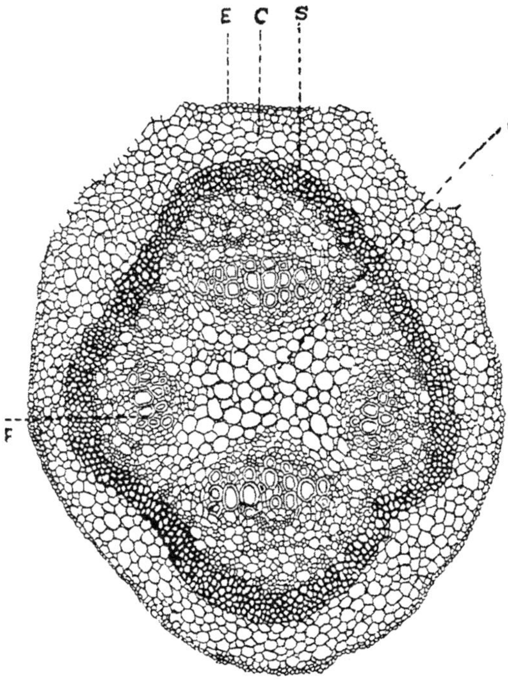

FIG. 8. Transverse Section of Petiole of Mimosa showing the four main bundles enclosed in protecting cylinder of sclerenchyma (two stray bundles at upper corners not shown).

E, epidermis; C, cortex; S, sclerenchyma; F, one of the four main vascular bundles; O, pith.

D

The results of the method of the removal of the various tissues are always brought forward in support of the view that the xylem is the conductor of impulse. The tabular statement given above shows how minute are the thicknesses of the different layers; there is no distance between them, for they are continuous. It is only under microscopic examination that it is at all possible to discriminate where one tissue ends and the other begins. It is therefore practically impossible to remove by hand (without the microscope to guide it) any particular layer, leaving the others uninjured. Again, if conduction persists after removal of the outer phloem, it does not at all follow that the xylem is the transmitting tissue, for the inner phloem would still be left for conduction. The method of removal of various tissues for the determination of the transmitting tissue is not only crude and beset with numerous sources of error, but the inferences drawn from the results contradict each other. Thus while some observers came to the conclusion that the xylem was the transmitting tissue, Haberlandt arrived at the definite opinion that the phloem, and not the xylem, was the conducting tissue. The only certain method for localising the conducting tissue is the electrical, and the results of this point definitely to the phloem as being the conductor of excitation.

Two Phloem-Strands, External and Internal

Objection has been raised to my assertion of the existence of an inner phloem; it has been said that, unlike the external phloem, the inner tissue consists only of parenchyma-cells, and that it contains no tube-cells. The above objection is entirely groundless, as will be seen from what follows, where I show how by means of selective staining it is possible to distinguish two neighbouring systems of tissue having different functions, or to establish the similar functional activities of two tissues which happen to be separated from each other.

The first step is to distinguish the two phloems from the other tissues. It is now recognised that the outer phloem is a conductor of excitation. In order to prove that there is an internal conducting phloem, it is necessary to establish the following points :

(i) That the internal phloem gives the same staining reactions as does the external; moreover, the reaction of the phloems, both external and internal, should distinguish them from the interposed xylem, and also from other adjacent tissues such as parenchyma of cortex and of pith.

(ii) That since the conductivity of the external phloem is associated with the presence of tubular cells, if the internal phloem is conducting, it must be characterised by similar tubular cells.

(iii) That the internal phloem does conduct excitation.

Differentiation of the tissues.—Differential staining shows the distinction in a striking manner. In order to secure a strong contrast, it is necessary to regulate the duration of application, for too prolonged action causes diffuse staining. By exact timing of the application, hæmatoxylin and safranin produce a very marked contrast between the xylem and the phloem. While the interposed xylem is stained red, the two phloems, external and internal, are stained deep violet. The other tissues are but faintly coloured.

In another method the section is first subjected for a short time to the action of hæmatoxylin, and is then treated for a longer period with Bismark brown. Examination of the transverse section shows that while the xylem, the parenchyma of the cortex, and the pith are coloured brown, the external phloem *p* and a tissue on the inner side of the xylem, which is in fact the internal phloem *p'*, are stained a deep violet (*cf.* fig. 9). Thus the staining reaction of the external and internal phloem is not only the same but is characteristically different from those of the neighbouring tissues.

Microscopic examination of the longitudinal section.—In order to determine the anatomical characteristics of the two phloems, a longitudinal section of one of the bundles

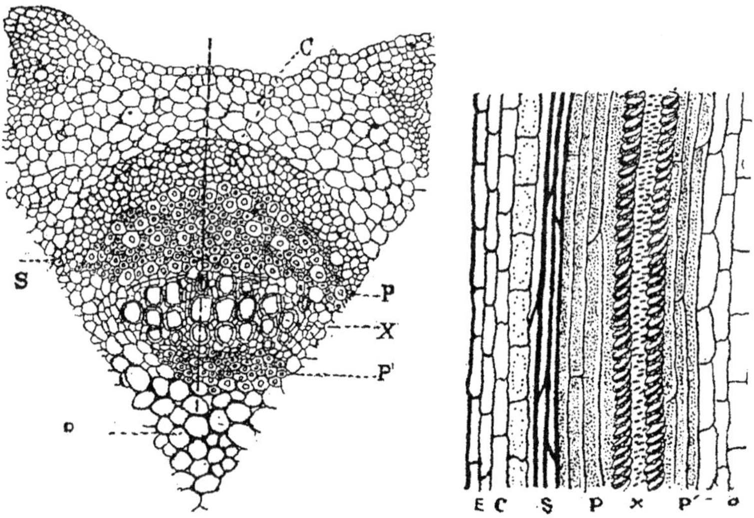

FIG. 9. Transverse and Longitudinal Sections of a Single Vascular Bundle.

Left figure : Transverse section. The dotted vertical line indicates the passage of the Electric Probe.
C, cortex ; S, sclerenchyma ; P, external phloem ; X, xylem ; P', internal phloem ; O, pith.
Figure to right : Longitudinal section of the bundle. Note elongated tubular cells, both in the external and in the internal phloem. (The section passed through one side of the bundle and not through the middle.)

was made. This is shown under high microscopic magnification (fig. 9), in which E is the epidermis ; C, parenchyma of the cortex ; S, sheath of protective sclerenchyma ; P, elongated tube-cells of the external phloem ; X, the vessels of the xylem ; P', tube-cells of inner phloem ; and O, the pith. The tubular cells in the inner phloem are as definite and distinct as those in the outer phloem. Relatively few of the tubular cells of the outer phloem have perforated septa, *i.e.* are sieve-tubes,[1] the other cells being imperforate, as are also the septa of the tubular cells of the inner phloem. But it must not be assumed that protoplasmic continuity is necessary for conduction of excitation, for I shall explain in the next chapter that excitation is transmitted across synaptic membranes in the animal nerve, and that similar transmission of excitation takes place in the plant across the septa of the conducting cells.

The fact that the inner phloem does actually conduct excitation is proved by my investigation on the localisation of the conducting nerve in the petiole by means of the Electric Probe, previously referred to, which will be fully described in Chapter X.

The line of passage of the Probe is indicated by the dotted vertical line in fig. 9. Galvanometric negativity, due to transmitted excitation, was found to occur only when the Probe came in contact with the phloem; the cortex, the xylem, and the pith did not show this characteristic reaction. The electric excitation was found to exhibit two maxima, one inside and the other outside the xylem. It was in fact this which led to the unexpected discovery of the two conducting phloem-strands, the bundle being thus bi-collateral.

The second physiological method of demonstrating the two conducting phloems is the detection of two definite excitatory impulses generated by separate stimulation of the external and internal phloems, fully described in Chapter XVI.

I may digress, for a moment, to point out that the presence of two conducting phloems in the bundles would appear to afford a satisfactory explanation of certain anomalous results obtained by Pfeffer in his experiments on the effect of an anæsthetic, chloroform, on conduction in the petiole. This method is not only crude but also

[1] Sieve-tubes would appear to have a physiological function other than that of conduction.

very uncertain in its action. It has already been pointed out that it is extremely doubtful if external application of chloroform would effectively paralyse the nerve in the interior of the petiole. It is, however, conceivable that after prolonged application a small quantity of the narcotic might, by absorption, get across to the outer phloem ; the

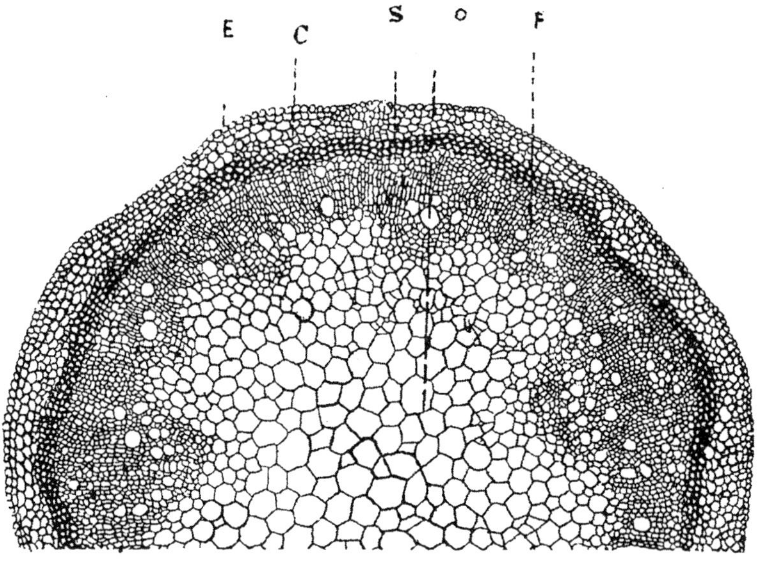

FIG. 10. Transverse Section of the Stem.
(One half shown in the figure.)

xylem would, however, interpose an additional barrier to the passage of the narcotic to the internal phloem. In such a case there would be paralysis of conduction in the outer phloem, while the inner phloem would still function in conducting excitation. Now it is obvious that, while a superficial mechanical stimulus will initiate an excitatory impulse in the outer phloem, a deep wound or cut is necessary to cause excitation of the inner phloem. The above considerations will probably explain Pfeffer's observation that while the excitation induced by a deep wound-stimulus was always conducted across the superficially narcotised area, the transmission of the excitation induced by mechanical stimulus applied on the surface was often blocked during transit through the chloroformed region. The anæsthetic reached and paralysed the external phloem, but failed to reach the internal phloem.

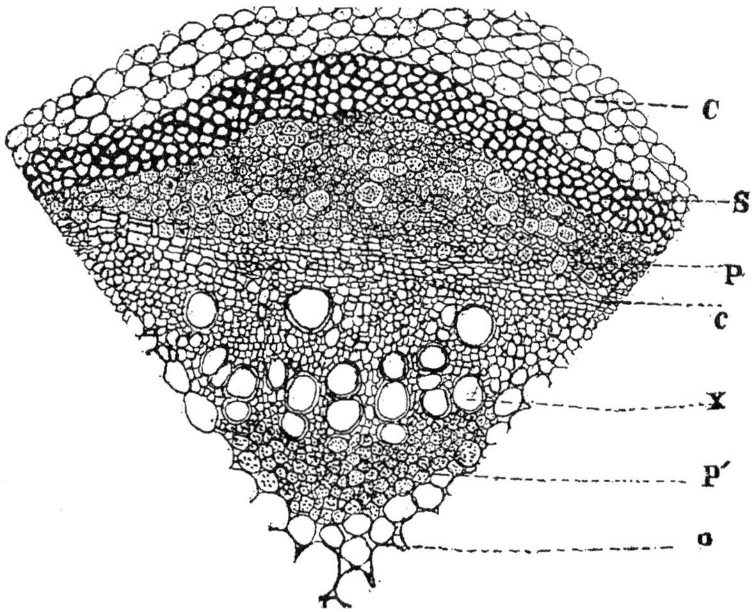

FIG. 11. Transverse Section of a Single Bundle in the Stem. C, cortex ; S, sclerenchyma ; P, external phloem ; *c*, layers of cambium ; X, xylem ; P′, internal phloem ; O, pith.

CONDUCTING TISSUE IN THE STEM

The supposition that the conducting mechanism in the stem is different from that of the petiole is altogether groundless. Fig. 10 shows the circular arrangement of a number of bundles in the stem, each bundle being laterally contiguous to the next. A magnified micro-photograph of one of the main bundles is given in fig. 11. The different tissues are essentially the same as those in the petiole: C is the cortex, S the sclerenchyma, P the external phloem, P′ the internal phloem, and O the pith. In the petiole, growth in diameter is completed within a short time, and little trace of a cambium is seen in the section; but in the stem the cambium-layers *c* are conspicuous.

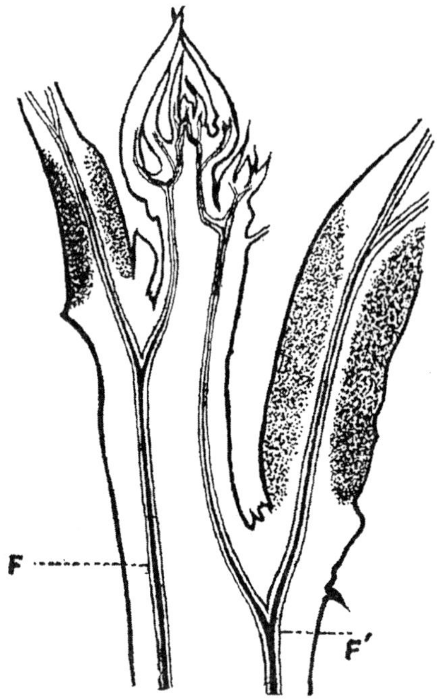

FIG. 12. Conducting Continuity between Stem and Leaves.

Longitudinal section of stem bearing leaves on each side. The two ascending bundles FF′ give lateral branches to leaves, and meet at apex. The double phloem stained violet stands out against the background. Pulvinus of leaves shaded.

NERVE-CONNECTION BETWEEN STEM AND LEAVES

There are two opposite main bundles in the stem which give off lateral branches to the odd and even series of leaves. A vertical section of a young stem in the plane of the leaf-insertions (fig. 12) shows this in a very clear manner. After treatment of the section with hæmatoxylin and safranin, the inner and outer phloems of each bundle appeared separately as two violet strands. The phloem-strands give off lateral branches to the leaves, thereby assuring conducting continuity between the stem and leaves. An impulse initiated by stimulation of the stem can thus be propagated in a centrifugal direction to the leaves; and an impulse generated in the leaves can, on the other hand, travel centripetally to the stem, and be then conducted up and down so as to cause the fall of the leaves. It will also be noted that the two main strands of

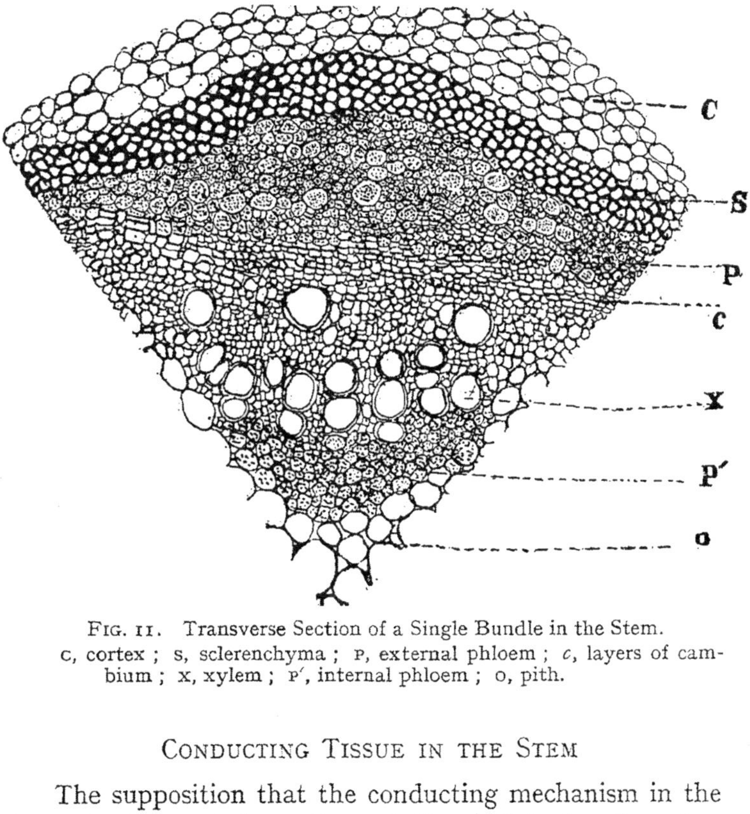

conducting phloem converge and meet at the apex of the stem. The establishment of this continuity explains how it is that, under moderately strong unilateral stimulus, the ascending impulse crosses over at the apex and becomes reversed into a descending impulse on the opposite side. (Experiment 3.)

Forcing the Lateral 'Block' by Stronger Stimulus

It has been pointed out that the continuity of the conducting phloem explains the unilateral propagation of excitatory impulse along the two main conducting strands. Inspection of fig. 10 makes it also clear that besides these main strands there are intermediate ones. The conducting phloems of the latter are not *continuous with, but contiguous to those of the two main strands.* There is thus a 'block' or resistance to the spread of excitation in a lateral direction; hence unilateral excitatory impulse is conducted lengthwise along the line of least resistance and not laterally across the stem.

Experiment 18. *Diffuse excitation under stronger stimulus.*—The resistance to the spread of excitation can however be overcome, as in the animal conducting tissue, by the application of a stronger stimulus. I repeated the experiment with all the specimens in which unilateral propagation of excitation occurred under diverse modes of moderate stimulation—by polar action of constant current, by scratch-stimulus and by induction-shocks. In the present series of experiments the stimuli were appropriately increased in intensity, with the result that the excitation, which had hitherto remained unilateral, now became outspread, causing responsive fall of all the leaves on *both* sides of the stem.

SUMMARY

Nervous impulse induces no visible change in the conducting tissue. The only means of its detection is by the concomitant electrical change of galvanometric negativity.

The tissue transmitting excitation in the plant has been definitely localised by means of the Electric Probe. The conducting tissue is the phloem.

Microscopic examination shows that the four main bundles of the petiole are separate from each other. They converge at the pulvinus and form an almost continuous ring.

From the results of electrical method of exploration, it is found that there are two conducting phloems in each bundle, one external and the other internal to the xylem. The bundle is therefore bi-collateral.

Microscopic examination confirms the existence of the two phloems, which stain deep violet with hæmatoxylin.

Tubular cells, which are characteristic of conducting tissue, are the main constituents of both the outer and the inner phloems.

The anatomical characteristics of the conducting tissue are the same in the stem as in the petiole. There are two main bundles containing conducting phloem on opposite sides of the stem, which run vertically up and down. The unilateral propagation of excitation under moderate stimulus is due to this particular nerve-distribution.

Microscopic examination shows that the nerve-strands of the two bundles meet at the apex. Hence it is possible for the ascending impulse on one side to cross over at the apex and to be conducted downwards on the other.

Anatomical examination shows that there is a continuity of nerve-connection between the stem and the leaf. Hence stimulation of the stem gives rise to a centrifugal impulse which causes the fall of leaves and the closure of leaflets ; conversely, strong stimulation of a sub-petiole gives rise to a centripetal impulse which, reaching the stem, causes fall of its leaves.

In addition to the two main bundles in the stem there are intermediate ones. There is no continuity, but contiguity, of the phloems in a lateral direction. A block or resistance to the lateral spread of excitation therefore exists. This block can, however, be overcome by a strong stimulus, when the excitation becomes outspread, causing fall of leaves on both sides of the stem.

CHAPTER V

TRANSMISSION OF EXCITATION ACROSS A SYNAPTIC MEMBRANE

THERE are reasons for the belief that in the animal there is no protoplasmic continuity across the nerve-junction where neurone joins neurone, the separating membrane being known as a synapse. The effect of discontinuity is most strikingly exhibited at the neuro-muscular junction where the intervention of the end-plate between the nerve and the muscle gives rise to a valve-action, on account of which excitation passes only in one direction. The conduction is therefore irreciprocal; for while there is no obstacle to prevent the passing of the excitatory impulse from the nerve into the muscle, the excitation of the muscle does not pass backwards into the nerve.[1]

Can a similar condition be discovered in Mimosa at the junction of the conducting nervous strands with the motor organ, the pulvinus? Taking the case of one of these strands which terminates centrally in the lower quadrant of the pulvinus, while at the periphery it is led to the second sub-petiole bearing the sensitive leaflets, stimulation of the second sub-petiole gives rise to an impulse which is propagated along the particular conducting strand in the petiole from the periphery to the centre. The impulse conducted across the junction between the conducting and the motor tissue causes contraction of the lower half

[1] 'There is a membrane intervening between the nerve-ending and the muscle-fibre supplied by it, as also between one neurone and the fibre connecting it with another neurone. The membrane is called by Sherrington "synaptic membrane" '—Bayliss—*Principle of General Physiology*, p. 434.

of the pulvinus and the resulting fall of the leaf. The impulse can thus pass easily from the conducting nerve to the contractile tissue. If now the contractile tissue of the pulvinus were directly stimulated, would the excitation pass in the opposite direction, from the excited lower half of the pulvinus to the nerve-end imbedded in it? If this occurred, the fact would be demonstrated by a centrifugal impulse, which on reaching the second sub-petiole would cause successive closure of its leaflets.

Experiment 19. *Effect of stimulation of the contractile cells.*—The contractile cells of the pulvinus can be stimulated without irritating the imbedded nerve. For this purpose the petiole is held within two clamping jaws, with padding of soft cotton-wool. The lower half of the pulvinus is now scratched superficially with a pin. The result of this is the contraction of the pulvinus, as evidenced by the fall of the leaf as soon as the jaws are slightly opened. The object of slightly opening the jaws is to prevent the impulsive fall of the leaf which might cause diffuse stimulation. With these precautions it was invariably found in these experiments that stimulation of the pulvinus is not followed by closure of the leaflets of any one of the sub-petioles; hence no conduction of excitation takes place from the contractile motor cells to the nerve. *The conduction at the junction of the nervous and contractile tissues in the pulvinus of Mimosa is therefore irreciprocal as at the neuro-muscular junction in the animal.*

The interposition of synapses in the path of nervous impulse in the animal also renders the conduction irreciprocal or unequal in the two directions. Two other modifications occur in the velocity of the impulse due to the presence of a synapse or of a series of synapses.

'If the passage (of the impulse) be too often repeated, phenomena of fatigue are produced and there is an increase of the block at each synapse. If, however, the stimulus be not excessive and the reaction not too frequently evoked the effect of passage of an impulse is to diminish the resistance, so that a second application of stimulus evokes the reaction more easily.'[1]

The above-mentioned characteristic tests may be applied, in determining the presence of a synapsoidal membrane in the conducting path in Mimosa, to ascertain

(1) Whether the conduction along the nerve is irreciprocal;
(2) Whether excessive stimulation induces fatigue;
(3) Whether a 'Bahnung' or Facilitation is produced as an after-effect of moderate stimulation.

The conducting tissue of Mimosa has been shown (see fig. 9) to consist essentially of elongated tubular cells, the transverse septa of which are mostly not perforated. The septa may therefore be regarded as synapsoids, if it is found that the transmission of excitation across them exhibits the characteristics already described.

Irreciprocal Conduction

This is demonstrated by the fact that the facility for conduction of excitation in the plant-nerve is unequal in the two directions. This statement holds good not only for *Mimosa pudica* but also for other sensitive plants. The intensity of stimulus employed was minimal.

Experiment 20. *Stem of Mimosa pudica.*—The intensity of stimulus unilaterally applied was gradually increased till it became minimally effective. Excitation was now found to be transmitted upwards and not downwards. Conduction under minimal stimulus is therefore irreciprocal.

Experiment 21. *Petiole of M. Spegazzinii.*—The stimulus of equi-alternating induction-shocks was applied below the pair of sub-petioles numbered 4 (fig. 13). The stimulus became minimally effective at intensity 1; transmission took place only in the centrifugal direction, as shown by the successive fall of the two pairs of sub-petioles 4 and 3,

[1] Starling—*Principles of Human Physiology* (1920), p. 305.

the leaflets of which also underwent closure. The distance of propagation of impulse was 37 mm., the transmission-time being 22 seconds; the velocity of transmission was

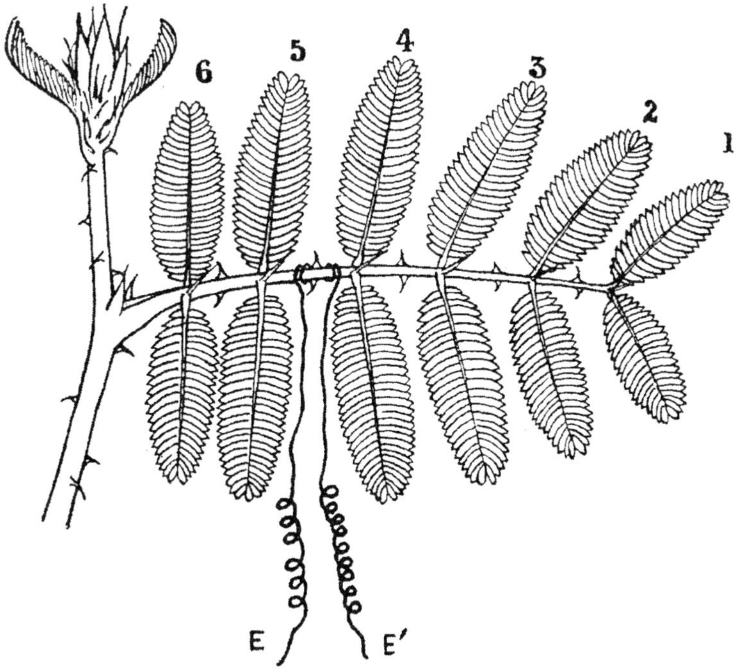

FIG. 13. Transmission of Excitation under Minimal and Maximal Stimuli.

Induction-shock applied by electrodes E, E′. Under minimal stimulus, transmission occurs only in the centrifugal direction. Under strong stimulus, excitation transmitted in both directions, the velocity in centrifugal direction being much greater than in the centripetal. (*M. Spegazzinii.*)

therefore 1·7 mm. per second. There was no evidence of transmission in the centripetal direction.

Experiment 22. *Averrhoa Carambola.*—The two electrodes for excitation by polar action of a constant current were applied midway on the petiole bearing numerous sensitive leaflets. Application of an E.M.F. of 6 volts was found to be minimally effective, giving rise to an impulse which was propagated in an outward or centrifugal direction, the three pairs of the leaflets to the right exhibiting closure in serial succession. There was no evidence of transmission in an inward or centripetal direction, the leaflets to the left remaining unexcited.

Experiment 23. *Biophytum sensitivum.*—The two electrodes of an induction-coil were applied midway on the petiole. As the secondary coil was gradually brought nearer the primary, a position was found for the minimally effective stimulus. On account of the partial block to the transmission of impulse in an ingoing or centripetal direction, the excitatory impulse was propagated only in the centrifugal direction.

The typical experiments carried out with diverse sensitive plants described above, prove that conduction is irreciprocal. They also indicate the existence of synapsoidal membranes which, by their valve-like action, permit propagation of impulse in one direction only.

Continuity between Irreciprocal and Preferential Conduction

It has been shown (Experiment 18) that the block or resistance to the passage of an impulse may be overcome by increasing the intensity of the stimulus. Excitation by unilateral stimulus which, when of moderate intensity, was only conducted lengthwise, was transmitted crosswise when the stimulus was stronger. It might therefore be expected that a hitherto ineffective transmission downwards in stems and inwards in petioles might become effective under increased intensity of stimulus. Irreciprocal conduction would thus be gradually transformed into conduction in both directions. Even after the partial forcing of the block, the valve-action of the synapsoid might still be exhibited by preferential conduction, in which case the velocity would be higher in the centrifugal direction than in the centripetal. These anticipations have been fulfilled in the results described below, in which the intensity of stimulus was increased from minimal to sub-maximal.

Stem of Mimosa pudica.—I have shown (Experiment 1) that under a sufficiently strong stimulus conduction takes place both upwards and downwards. But the velocity in the upward direction was the greater; for while the transmission upwards through 37 mm. took place in 10 seconds, the transmission in the downward direction through 28 mm. required 20 seconds. The velocity in the former case was 3·7 mm. per second, while in the latter it was 1·4 mm. per second. Hence the velocity in the preferential upward direction was about 2·7 times that in the opposite direction.

The internodes below are older than those above, and the observed difference might have been partially due to the diminished conductivity of the older tissues. It is therefore desirable to observe the velocity in the two directions in the same internode. In Experiment 3, carried out with a highly excitable specimen, a moderately strong unilateral stimulus initiated an impulse which was at first conducted upwards; it then crossed to the opposite side and was propagated downwards along the same internodes. A large number of experiments carried out in this manner showed that the rate of propagation upwards was from 3 to 3·5 times quicker than that in the downward direction.

The synapsoidal block is more and more effectively forced under increasing intensity of stimulus. Thus, while under minimal stimulus, the conduction is irreciprocal, under sub-maximal stimulus a partial forcing of the block occurs, the conduction in the preferential direction being considerably quicker than in the opposite. Under maximal stimulus the forcing of the block is more complete, and the two velocities then tend to become nearly equal.

Experiment 24. *Petiole of M. Spegazzinii.*—In Experiment 21 a stimulus-intensity of 1 caused the fall of the outer sub-petioles 4 and 3 (fig. 13). Using the same specimen, the stimulus was increased to 1·5; this gave rise to a stronger impulse which caused the fall not only of sub-petioles 4 and 3 but also of sub-petioles 2 and 1. The excitation under minimal stimulus had not been at all transmitted inwards; but under the stronger stimulus of 1·5 the impulse also travelled in a centripetal direction and caused the fall of the sub-petiole 5, below the point of application of stimulus. But this occurred five seconds after the fall of the outermost pair of sub-petioles. The rate of conduction was therefore much quicker in the centrifugal direction than in the centripetal.

Experiment 25. *Petiole of Averrhoa.*—The intensity of stimulus was increased slightly above the value for irreciprocal conduction. The centrifugal impulse caused closure of all the leaflets to the right, but the centripetal impulse reached only one pair of leaflets to the left. Even in this case of transmission in both directions, the velocity in the centrifugal direction was about three times the greater.

Experiment 26. *Biophytum.*—Results in every way similar to the above were obtained with this plant, which is far more sensitive than Averrhoa. When the intensity of stimulus was slightly increased above the minimal, the impulse was transmitted in both directions; but while all the leaflets to the right (centrifugal) closed, only a few of the leaflets to the left (centripetal) underwent closure. A still stronger stimulus caused a more effective conduction in both directions, but the centrifugal velocity was one and a half times greater than the centripetal.

Irreciprocal conduction and conduction in a preferential direction both indicate the existence in the conducting cells of synapsoidal membranes across which the excitation is transmitted with greater or less facility. Their existence is, as previously stated, also indicated by the fatigue of conduction under excessive stimulation, and by 'Bahnung' or Facilitation as an after-effect of moderate stimulation. I proceed to give an account of the effects of excessive and moderate stimulation on the velocity of conduction which I was able to determine with the highest degree of accuracy by the automatic method fully described in the next chapter.

FATIGUE OF CONDUCTION

The experiments were carried out with the petiole of *Mimosa pudica*. Fatigue was produced by the previous application of an excessively strong stimulus. It is to be noted that the experiments were carried out with vigorous specimens in which the normal velocity was high.

Experiment 27.—In a particular specimen the normal velocity was found to be 18·7 mm. per second. In order to observe the after-effect of excessive stimulation, the end of the petiole beyond the point of application of the testing stimulus was then cut off; after recovery, the record of velocity of transmission under the test-stimulus was taken once more. It was found that the excessive wound-stimulus had induced a depression of the conducting power, the velocity being reduced from the normal 18·7 to 10·7 mm. per second.

In another experiment the normal rate of 30 mm. was reduced by fatigue to 19 mm. per second.

'BAHNUNG' OR FACILITATION

In highly excitable specimens of Mimosa, strong and long continued stimulation induces, as stated above, a depression or fatigue in the rate of conduction. But in less excitable specimens the after-effect of stimulus is to confer an enhanced power of conduction, as if the passage of impulse removed some resistance or block that existed before.

Experiment 28.—The conducting power of the specimen was so low that the impulse due to the test-stimulus applied on the petiole, at a distance of 15 mm. from the pulvinus, failed to be transmitted. On application of a much stronger stimulus the excitatory impulse was found to be effectively transmitted with a considerable speed. After a period of rest the original ineffective stimulus was applied once more. The impulse was now found to be transmitted, the velocity being as high as 25 mm. per second. This enhanced conducting power began slowly to decline, and after an hour the velocity was reduced to 4 mm. per second. The application of a strong stimulus was once more found to confer enhanced conducting power, the velocity under the test-stimulus being increased from 4 to 25 mm. per second. *The conducting path thus becomes canalised by stimulation.*

Experiment 29. *Petiole of M. Spegazzinii.*—A similar effect was strikingly demonstrated by this specimen, which was only moderately excitable and conducting. A stimulus of intensity of 1·5 from an induction-coil, applied below sub-petiole 3, failed to initiate any excitatory impulse. Stimulus of intensity 2 was similarly ineffective; and it was only under an intensity of 2·5 that the impulse was effectively transmitted, causing the fall of sub-petioles 3, 2, and 1 (see fig. 13). After this the formerly ineffective stimulus of 1·5 was applied once more, and was now found to be effectively transmitted, causing the fall of all the sub-petioles.

Summary

Conduction of excitation takes place by means of the tubular cells in the phloem. The transverse septum between any two of the cells acts as a synapsoidal membrane.

The characteristics of conduction across the synapsoidal membrane in the plant are similar to those across the synaptic membrane in the animal.

Conduction is irreciprocal at the neuro-muscular junction. It is also irreciprocal at the junction of the plant-nerve with the contractile tissue of the pulvinus.

Under minimal stimulus the conduction in the plant-nerve is irreciprocal, excitation being easily transmitted in one direction and not transmitted in the other.

Under sub-maximal stimulus there is a partial forcing of the block, with the result of transmission in both directions: the irreciprocal conduction becomes transformed into preferential conduction, the velocity being greater in one direction than in the opposite.

Fatigue in conduction takes place after excessive and long-continued stimulation.

The phenomenon of 'Bahnung' or Facilitation occurs as an after-effect of short-lived stimulation. A previously ineffective stimulus now becomes effective. Stimulus thus canalises its own conducting path.

The characteristics of nerve-conduction in the plant are similar to those in the animal.

CHAPTER VI

AUTOMATIC RECORD OF VELOCITY OF TRANSMISSION IN PLANTS

THE results of the qualitative experiments described in the last chapters are sufficiently convincing to prove the nervous character of the transmission of impulse in plants. But for quantitative research it is necessary to devise a method for the determination of its velocity of as high a degree of accuracy as that attained in the determination of the velocity of nervous impulse in animals. For this purpose two conditions are essential: (1) a device for stimulation such that the intensity may be kept constant or be varied in a graduated manner; (2) an automatic recorder by which a time-interval, shorter than a hundredth of a second, may be measured with the utmost accuracy.

QUANTITATIVE STIMULATION

The excitatory impulse may be initiated in various ways, of which scratch-stimulation is an example. Section of the petiole is also very effective, though a hydro-mechanical disturbance is produced along with the excitatory impulse. The application of a hot glowing point causes an intense excitation which is transmitted to a great distance. Stimulation by a drop of hydrochloric acid is equally effective.

The above modes of stimulation may be employed for purposes more or less qualitative. They cannot, however, be repeated with equal intensity in successive experiments, nor can they be gradually increased from minimal to maximal.

The most satisfactory mode of stimulation of the plant to obtain mechanical response is the electrical. It can be readily effected by electric shocks from an induction-coil, the effective intensity of which is determined jointly by the intensity of the shock and the duration of its application. This latter can be very accurately adjusted by means of a Plunger actuated by clockwork, which completes the electric circuit of the primary coil. The Plunger dips into a deep cup of mercury; the duration of the closure of the current may be adjusted from half a second to five seconds. This is done by raising or lowering the mercury cup by a rack and pinion. It is better to use only a moderate intensity of induction-current, the effectiveness of which can be increased by prolonging the duration of application; under these conditions the excitability of the tissue remains unchanged for a comparatively lengthy period.

THE RESONANT RECORDER

The most difficult problem is the exact determination of the time of transmission of impulse as indicated by observation of the responsive fall of the leaf.

The leaf attached to a writing-lever may be made to record its responsive movement on a moving recording plate: the lever bears the writer at right angles to it. The recording plate is raised, and then allowed to descend: during its fall a brief contact is made which completes the electric circuit of the primary coil P (fig. 14), and generates a current in the secondary coil the terminals of which are applied on the petiole at A, a definite distance from the motile pulvinus B. The precise moment of stimulation is marked on the record as an arrow. The response is not immediate, but there is a definite interval between the application of stimulus and the beginning of response. This interval T is made up of the time t required for excitation to travel along the definite length of the petiole, *plus* the time required for starting the motile mechanism into action, which is the latent period L of the pulvinus. For the exact determination of the true transmission-time t, the latent period L has to be subtracted from the observed interval T,

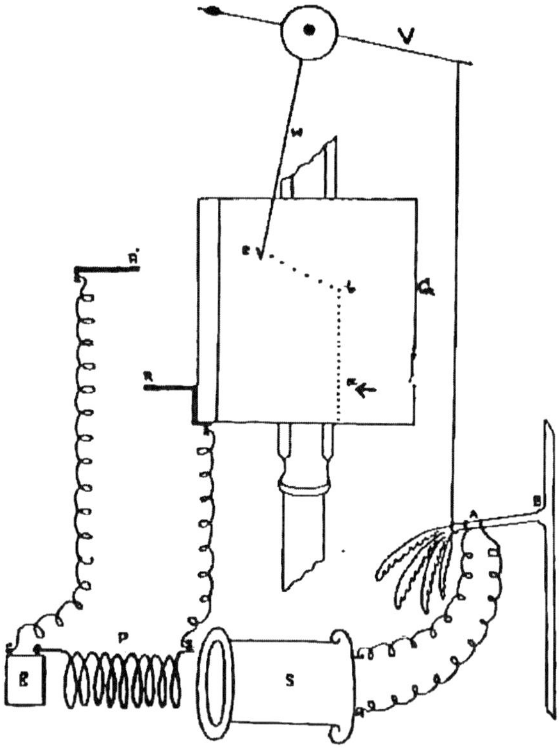

FIG. 14. Diagrammatic representation of the Recorder.

V, one arm of lever attached to leaf ; W, the writer. The falling plate during descent makes electric contact of R with R′ causing induction-shock by the secondary coil S. Stimulus applied at *a* causes response later, at *b*.

between stimulation and response. A highly sensitive method has therefore to be devised for measurements of time-intervals shorter than a hundredth of a second.

All difficulties of recording and measuring of short intervals of time are overcome by my Resonant Recorder (fig. 15), in which the writer is not in continuous contact with the recording plate, but makes regular intermittent contacts with it so that the error due to friction is completely removed. The writer, consisting of a thin steel wire, is

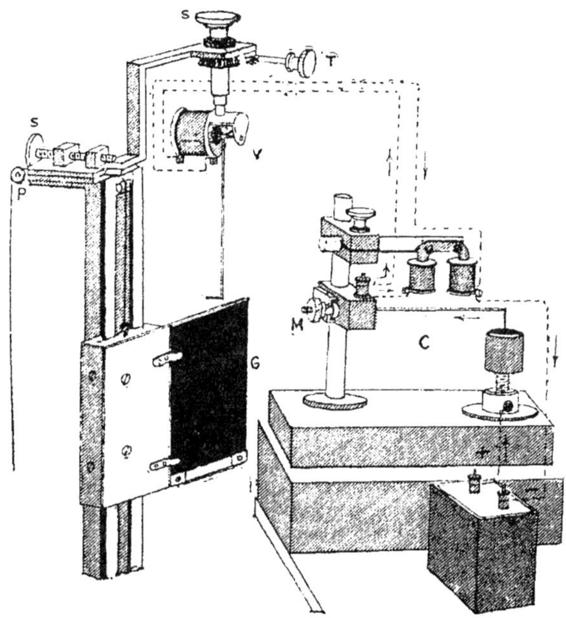

FIG. 15. Upper part of Resonant Recorder. (From a photograph.)

Thread from clock, not shown, passes over pulley P, letting down recording-plate; S′, screw for adjusting distance of writing-point from plate; S, screw for vertical adjustment; T, tangent-screw for exact adjustment of plane of movement of recorder, parallel to writing-surface; V, axis of writer supported perpendicularly at centre of circular end of magnet; C, coercer; M, micrometer-screw for adjustment of length of coercer; G, smoked-glass plate.

supported at the centre of one pole of an electro-magnet, which is periodically magnetised by the passage of an interrupted current. The interruption is produced by the vibrating spring C, which is designated the Coercer (fig. 15), the period of vibration of which can be accurately adjusted. Different writers are tuned exactly to vibrate 10, 100 or 200 times per second. When the Coercer is adjusted to these vibrations, the writer is thrown into sympathetic vibration. The record then consists of a series of dots;

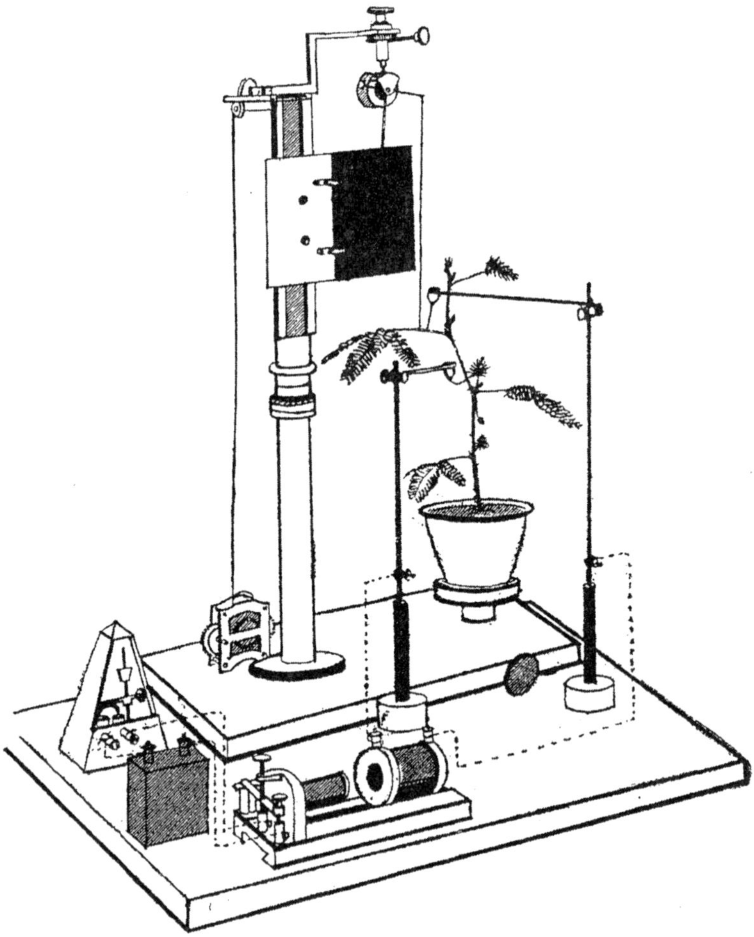

FIG. 16. Apparatus for determination of latent period and velocity of transmission of excitation in Mimosa.

when they are, say, a 200th of a second apart, the interval between successive dots in the record is ·005 second. It is not difficult to measure one-fifth of the distance between successive dots: so calculations can be carried to the thousandth part of a second. The intermittent contact serves two most important purposes. First, it eliminates all error arising from friction. Secondly, the distance between successive dots in the record itself enables measurement of extremely short time-intervals to be made with the highest degree of accuracy. The plant-record or phytogram is also its own chronogram.

Determination of the Latent Period

Fig. 16 gives the general arrangement of the apparatus for experiments on the determination of the latent period and of the velocity of transmission of excitation. For direct stimulation of the pulvinus, the two electrodes are

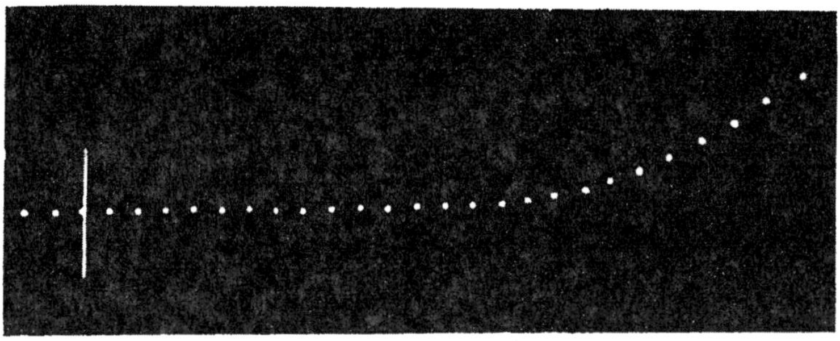

Fig. 17. Record of Latent Period of Mimosa, with 200 Vibration Recorder.

attached to the petiole and the stem respectively, the shock being passed through the included motile organ. The record of an experiment for the determination of the latent period is given in fig. 17. There are 15·2 spaces between the incidence of stimulus and response. The writer being tuned to 200 vibrations per second, the interval between the dots is ·005 second, and therefore the latent period of the specimen is ·076 of a second.

The value of the latent period depends on the season and on the physiological condition of the specimen. It varies in different specimens from ·05 to ·12 second. The approximate average value of L may be taken as ·10 second.

Experiment 30. *Effect of fatigue.*—If a sufficient period of rest be not allowed between two successive responses the latent period becomes prolonged, indicating the effect of fatigue. Thus in a particular experiment the normal latent period of ·10 second (upper record, fig. 18) was prolonged under fatigue to ·14 second (lower record). Under excessive fatigue the motile excitability becomes temporarily abolished.

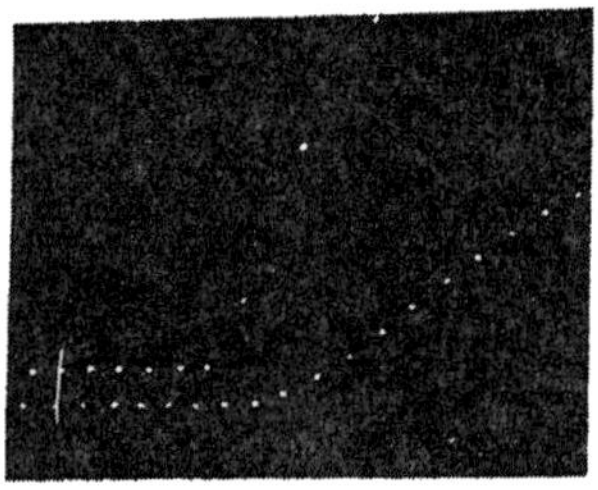

FIG. 18. Effect of Fatigue in prolonging the Latent Period: upper record, normal; lower record, fatigue; frequency of vibration 50 per second.

Experiment 31. *Effect of temperature.*—The latent period becomes prolonged under cold and shortened under a rise of temperature. The latent period of a particular specimen was ·14 second at 24° C.; at 29° C. it was shortened to ·10 second. At the higher temperature of 33° C., it was reduced to ·07 second.

DETERMINATION OF VELOCITY OF TRANSMISSION

(a) *Ordinary Method*

In order to determine the velocity of transmission induction-shock stimulus is applied to the petiole at a definite distance *d* from the responding pulvinus. In order to obtain the true transmission-time *t*, it is necessary, as previously explained, to subtract the latent period L from the observed total time T. The velocity of transmission is then found by dividing the distance by the true time. The necessary data are therefore the distance *d* between the stimulated point and the pulvinus, the total time T between the application of stimulus and the initiation of response, and the latent period L of the individual pulvinus.

In obtaining a record of response to indirect stimulation the two electrodes E and E′ are applied on the petiole, the proximal electrode being at a distance, say, of 30 mm. from the pulvinus. After taking one or more such records, an additional record is taken of the response to direct stimulation; this latter gives the latent period L of the particular specimen.

Experiment 32.—In the experiment I am about to describe the specimen of Mimosa was very vigorous. The distance at which the stimulus was applied was 30 mm. from the responding pulvinus, and the intensity of stimulus was 3 units, which was maintained constant in successive records The frequency of the vibrating-recorder was 10 per second; hence the distance between any two

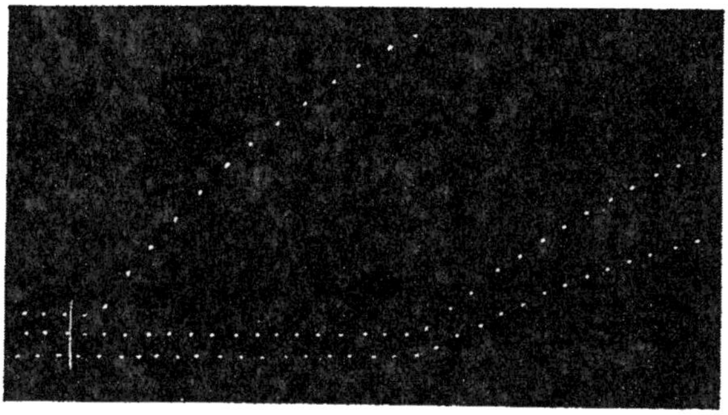

FIG. 19. Determination of Velocity of Transmission of Excitation in petiole of Mimosa.

Two lower records are in response to indirect stimulation applied at a distance of 30 mm.; upper record of response to direct stimulation gives the latent period. Recorder 10 V. per second.

successive dots in the record represents a time-interval of one-tenth of a second; from the record itself it is not difficult to estimate an interval of even one-fifth of that amount. The lowest of the three records (fig. 19) represents the results of the first experiment. It will be seen that the interval between the stimulation and the beginning of response is 16·2 spaces, each of the value of ·1 second. The total time T is therefore 1·62 seconds. After a suitable interval necessary for a complete recovery, a second record was taken, under the same conditions, on the same plate.

The minimum interval necessary for complete recovery varies from 15 to 20 minutes, depending on the condition of the specimen and the season. The total time in the second record is the same as before, namely, 1·62 seconds. The third record was taken under direct stimulation, and the latent period obtained therefrom is ·12 second. The velocity of transmission obtained from the two experiments is identical, namely:

$$t = 1{\cdot}62'' - {\cdot}12'' = 1{\cdot}5''$$
$$V = \frac{d}{t} = \frac{30}{1{\cdot}5} = 20 \text{ mm. per second.}$$

(b) *The Differential Method*

In order to put the constancy of these results to a still more rigorous test, I modified the experiment in the following way, employing the Differential Method. The stimulus was first applied at a distance d from the responding pulvinus, and the total time T was found from the record. In the next experiment the distance of the point of stimulation was reduced to d_1, and the corresponding total time T′ found in the usual manner. And lastly, a record was taken under direct stimulation for the determination of the latent period L.

These observations provide three different sets of data for the determination of the value of the velocity of transmission. By two of these is obtained the velocity in the usual manner from the distance, the total time, and the latent period. In the third, knowledge of L is not required, for $T - T_1$ gives the time of transmission through the distance $d - d_1$. Hence three separate determinations V_1, V_2, V_3 are obtained with the same specimen, viz.:

$$V_1 = \frac{d}{T - L};\ V_2 = \frac{d_1}{T_1 - L};\ V_3 = \frac{d - d_1}{T - T_1}$$

The rigour of the test of constancy will be gauged by the extent to which the determinations of V_1, V_2, and V_3 are consistent with one another.

Experiment 33.—In an experiment carried out in this

way the intensity of stimulus applied was 3 units. In the first experiment the point of application of the stimulus was at a distance of 30 mm.; the total time was found to be 1·9 seconds. In the next experiment the distance was reduced to half, *i.e.* to 15 mm., and the total time was found

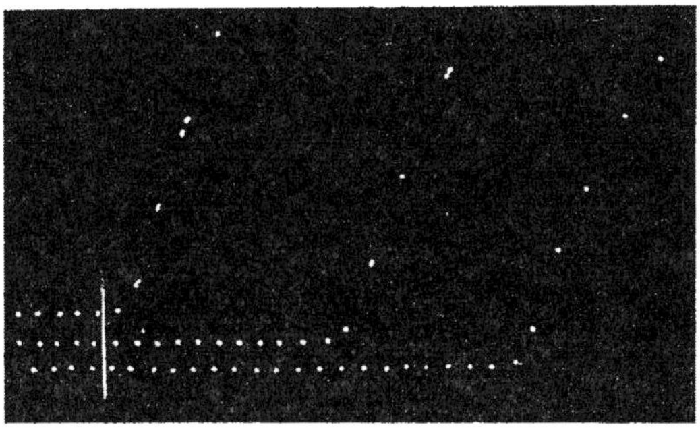

FIG. 20. Determination of Velocity by Differential Method. The records from below upwards are in response to stimuli applied at distances of 30 mm., 15 mm., and directly. Recorder 10 V. per sec.

to be 1 second. And lastly, the latent period, under direct stimulation, was found to be ·08 second (fig. 20). Thus—

$$V_1 = \frac{30}{1 \cdot 9 - \cdot 08} = 16 \cdot 4 \text{ mm. per second.}$$

$$V_2 = \frac{15}{1 - \cdot 08} = 16 \cdot 3 \text{ mm. per second.}$$

$$V_3 = \frac{15}{1 \cdot 9 - 1} = 16 \cdot 6 \text{ mm. per second.}$$

The three results thus obtained from independent data are seen to be extremely consistent. They bear very emphatic testimony not only to the accuracy of the method, but also to the constancy of velocity in a given specimen under unvarying external conditions.

The velocity of transmission is modified by the individual vigour of the specimen, by the temperature, and by the season. In summer the velocity in thick petioles is about 30 mm. per second; in winter it is as low as 5 mm. The velocity of nervous impulse in Frog's nerve is about

27,000 mm. per second, but in Anodon it is as low as 10 mm. per second.

The velocity of nervous impulse in a thick petiole of Mimosa is thus a thousand times less than in the Frog, but about three times greater than in a lower type of animal such as Anodon. I shall presently show that the velocity in *thin* petioles is very much higher than in thick specimens. The velocity in the stem and in the sub-petiole is, generally speaking, considerably lower than in the petiole, being of the order of 4 mm. per second.

VELOCITY OF TRANSMISSION IN THIN PETIOLES

Experiment 34.—The following results present many interesting features. I first determined the velocity of transmission in the thick petiole of a particular Mimosa. The day was very bright and dry and the temperature was at an optimum degree of 33° C. (April). On account of the exceptionally favourable conditions of light and temperature, the rate of conduction in the thick petiole was found to be as high as 55 mm. per second.

Lower down, the plant bore leaves of a different type, the petioles of which were extremely thin. I obtained a record of the velocity of transmission in a thin petiole, employing a recorder vibrating 20 times in a second; the successive spacings therefore represent ·05 second. The lower figure is the record of the transmitted excitation, the stimulated point on the petiole being at a distance of 20 mm. from the pulvinus. The upper record gives the latent period under direct stimulation (fig. 21). The data for the determination of the velocity are:

Latent period L = 1·5 spaces.

Total time between stimulus and response = 2·5 spaces.

True transmission-time through 20 mm. = 2·5 − 1·5 = 1 space = ·05 second.

$$\text{Velocity of transmission} = \frac{20}{\cdot 05} = 400 \text{ mm. per second.}$$

The above results are by no means exceptional, for in five other determinations of the velocity in thin petioles the rate was found to be very much higher than in thick petioles. The value varied in different specimens from

FIG. 21. Record of Velocity of Transmission of Excitation in Thin Petioles.

The lower record gives total time of transmission through 20 mm. Upper record gives the latent period. Vibration-frequency of recorder, 20 per second.

400 mm. to 120 mm. per second. The average velocity in thin petioles may therefore be taken as about nine times greater than that in thick petioles. An explanation of the difference of the velocity in thick and thin petioles will be given later (see Chapter XVI.).

AFTER-EFFECT OF STIMULATION ON TRANSMISSION

A remarkable fact noticed in quantitative determinations is the enhancement of velocity which results as an after-effect of stimulation. Thus stimulus applied for the first time gives rise to an impulse which is transmitted at a slow rate; successive stimulations enhance the rate still further, until an optimum rate is attained which remains constant for a considerable length of time. This is the *steady condition*, and the effects of external agencies on conduction can only be studied after the attainment of this condition. The enhancement of conducting power by a preliminary stimulation is evidence in support of the existence of synapsoidal membranes in the conducting tissue, the resistance of which is for a time overcome by the action of a stimulus.

Influence of Tonic Condition on Conductivity

Specimens of Mimosa are found to be in various conditions of physiological vigour. Some are in an optimum condition, others in an unfavourable sub-tonic condition. The velocity in the former is relatively high, while in the

FIG. 22. After-effect of Intense Stimulation in depressing rate of conduction in a normal specimen.
(1) record before, and (2) after injury. (Dot-intervals, 0·1 second.)

latter it is low. The after-effect of intense stimulation is diametrically opposite in the two cases, as will be seen in the following experiments.

Experiment 35. *After-effect of intense stimulation on normal specimens.*—Record (1) in fig. 22 gives the time of transmission through 15 mm. length of the petiole, the velocity of transmission being found to be 18·7 mm. per second. The end of the petiole beyond the point of application of the testing stimulus was then cut off, and record of velocity of transmission taken once more. It will be seen from record (2) that the excessive stimulation caused by the cut had induced a depression of the conducting power, the velocity being reduced to 10·7 mm. per second. Excessive stimulation of normal specimens thus temporarily depresses the conducting power.

Experiment 36. *After-effect of intense stimulation in sub-tonic specimens.*—I will now describe an experiment to show that an identical agent, on account of difference in the tonic condition of the tissue, gives rise to a diametrically opposite effect. I took a specimen in a sub-tonic condition, in which the conducting power of the tissue was so far below par that the test-stimulus applied at a distance of 15 mm. failed to be transmitted. The end of the petiole at a distance of 1 cm. beyond the point of application of the test-stimulus was now cut off. The after-effect of this injury was found so to enhance the conducting power that the excitation previously arrested was now effectively transmitted, the velocity being 25 mm. per second. This enhanced conducting power began slowly to decline, and after half an hour the velocity had declined to 4·1 mm. per second. The end of the petiole was cut off once more, and the effect of injury was again found to enhance the conducting power, the velocity of transmission being restored to 25 mm. per second.

SUMMARY

The latent period of the pulvinus may be determined with great accuracy by means of the Resonant Recorder, by which it is possible to estimate time-intervals as short as a thousandth part of a second.

The lowest value of the latent period of the pulvinus of Mimosa is ·06 second, the average value of the latent period being ·1 second.

Fatigue prolongs the latent period.

The latent period is shortened, within limits, by a rising temperature.

By applying stimulation of constant intensity, and by allowing proper intervals of rest, successive values of the velocity of transmission were obtained which were constant.

Consistent results are obtained by the employment of the Differential Method.

In thick petioles the velocity in summer is about 30 mm. per second, whereas in thin petioles the velocity may be as high as 400 mm. per second.

The velocity of transmission of excitation in the petiole of Mimosa is intermediate between the velocities of the nervous impulse in higher and lower animals.

The velocity in the stem and that in the sub-petiole of Mimosa are much lower than that in the petiole.

The tonic condition of a tissue has an influence on its conductivity and the induced variations of it. In a tissue in optimum condition the velocity is high, and excessive stimulation induces temporary depression of the conducting power. In a sub-tonic tissue the velocity of conduction is low, and the after-effect of intense stimulation is to initiate or enhance the rate of conduction: thus the conducting path is canalised by the stimulus.

CHAPTER VII

EFFECT OF EXTERNAL CHANGES ON THE VELOCITY OF TRANSMISSION

THE velocity of transmission has been shown to be constant under uniform external conditions. I will now describe the effects of external changes in inducing variations in the velocity of transmission in plants. For this purpose it is necessary to obtain two records in succession, the first under normal and the second under the changed condition. The difference reveals the effect of the changed condition on the velocity.

EFFECT OF DESICCATION

Experiment 37. *Effect of application of glycerine.*—It is known that desiccation, generally speaking, enhances the excitability of animal nerve. As glycerine, by absorption of water, causes partial desiccation, I tried its effect on conduction of excitation in the petiole of Mimosa. Enhancement of conducting power may be exhibited in two ways: either by an increase of the velocity of transmission; or, secondly, by an enhancement of the intensity of the transmitted excitation, which would be manifested in a greater amplitude of the response of the motile indicator. In fig. 23 are given two records, one before and the other after the application of glycerine on a length of petiole through which excitation was being transmitted. The time-records demonstrate the enhanced rate of transmission after the application of glycerine. The increased intensity of transmitted excitation is also shown in the enhanced amplitude of response exhibited by the more erect curve of the upper record.

EFFECT OF VARIATION OF TEMPERATURE

In animal nerve the velocity is increased under a rise of temperature. Conversely, under a falling temperature, the velocity is lowered, and conduction becomes blocked when the temperature is sufficiently low. I will show that precisely similar effects occur in the conduction of the excitatory impulse in Mimosa.

Experiment 38.—The plant was placed in a thermal chamber, the rise of temperature of which could be regulated

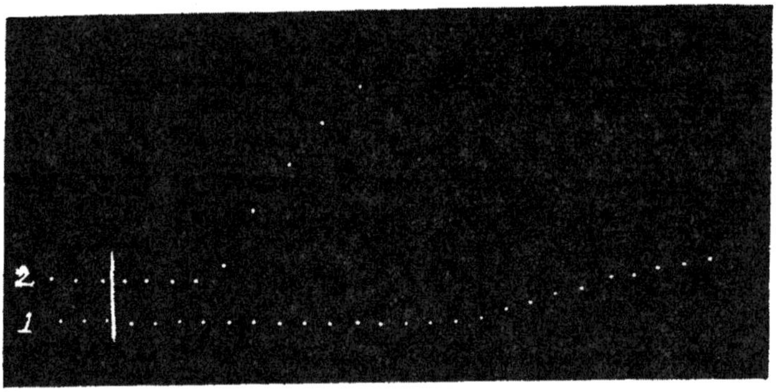

FIG. 23. Action of Glycerine in enhancing the speed and intensity of transmitted excitation.

Stimulus applied at the vertical line. Successive dots in record are at intervals of 0·1 second. 1, normal response.

by a suitable electric device. In fig. 24 time-records are given of the transmission of excitation at temperatures 22° C., 28° C. and 31° C. The experiment was carried out in winter in Calcutta, when the temperature of the room was 22° C., and the normal velocity, on account of the unfavourable season, was comparatively low. A maximal stimulus was applied at a distance of 10 mm. from the responding pulvinus. The lowest of the three records gives the transmission-time at temperature 22° C.; the middle record was taken at 28° C., and the uppermost one at 31° C. The records make it quite evident that the velocity continuously increased under a rising temperature. The total time (including the latent period) at 22° C. was 2·94 seconds; at 28° C. it was 1·69 seconds; and at 31° C. it was further shortened to 1·2 seconds. In a previous experiment the variation induced in the latent period by change of temperature had been determined; at 23° C. it was ·165 second; at 28° C. it was

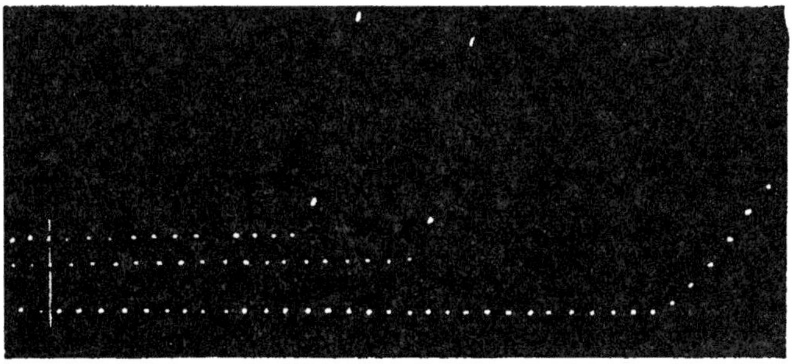

FIG. 24. Effect of Temperature in enhancing Velocity of Transmission.

The three records, from below upwards, are for temperatures of 22° C., 28° C., and 31° C. respectively.

·12 second; and at 33° C. it was further shortened to ·065 second. These variations are very slight as compared with the total period of transmission. After making the small corrections for the change in the latent period, the velocity of transmission at 22° C. was 3·6 mm. per second; at 28° C. it was 6·3 mm.; and at 31° C. it was 9 mm. per second. The results of numerous other experiments showed that the velocity was always increased by rise of temperature.

PHYSIOLOGICAL BLOCK

Transmission can be arrested by treating the conducting tissues in various ways; that is, a physiological block can be induced. The method employed is illustrated by fig. 25. The electric stimulation is applied at E and the various blocking agents are applied at B in the path of conduction.

Block by Local Cooling

The enhancement of conducting power by warmth has been demonstrated in the experiment just described. Conversely, conductivity is depressed by lowering the temperature, and this may become so great as to induce an actual arrest of conduction. The object of the present investigation being the determination of the influence of cold on conductivity, special care was taken that the

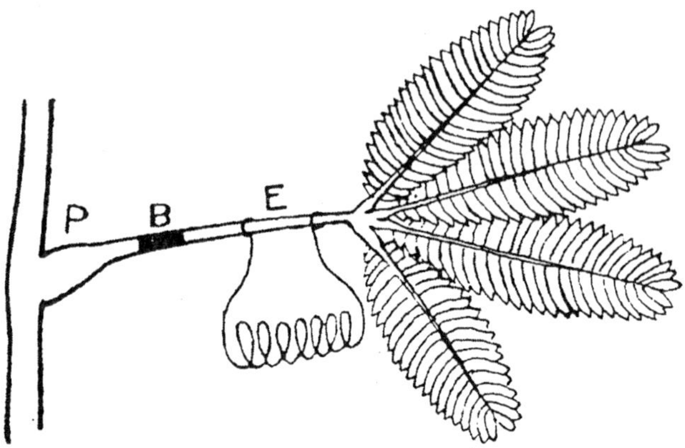

FIG. 25. The various physiological blocks are interposed at B, in the path of conduction.

lowering of temperature did not in any way affect either the receptive excitability of the stimulated point, or the motor excitability of the responding pulvinus. In order to ensure this, cold was applied locally at the middle of the petiole. The experiment was carried out in summer, the specimen being moderately sensitive. A strip of cloth 10 mm. in breadth was wrapped round the petiole midway between the stimulated point and the pulvinus, and the local lowering of temperature was produced by application of cooled water; excessive lowering was secured by placing powdered ice on the piece of cloth. The stimulus employed was maximal.

Experiment 39.—Successive records were taken at

intervals of 20 minutes, which is more than sufficient time for complete recovery from previous stimulation. Record (1) gives the time-interval between stimulus and response under normal conditions (fig. 26). There are 20·5 intervening spaces, each representing ·1 second. The latent period is 1·5 seconds. The true time of transmission is thus 1·9 seconds for 30 mm.; the transmission-time through 10 mm. is therefore ·63 second.

The next record was taken when an intervening length of 10 mm. on the petiole was moderately lowered in tempera-

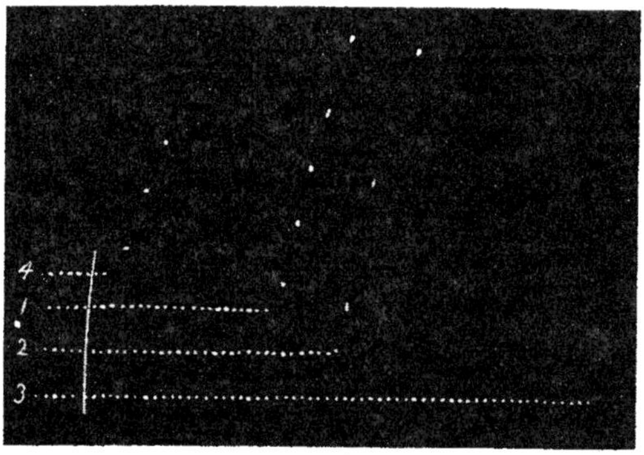

FIG. 26. Effect of Cold in inducing Retardation and Arrest of Transmission.

(1) Normal record; (2) Retardation due to slight cooling; (3) Arrest of conduction brought about by intense cold; (4) Record of direct stimulation.

ture by the application of cold water. This cooling should be commenced immediately after the previous responsive fall of the leaf. This not only gives sufficient time for localised cooling of the petiole, but also avoids the excitatory disturbance of the pulvinus caused by a sudden application of cold water to the petiole. During the localised cooling of the petiole, the leaf re-erects itself and becomes fully sensitive when the time arrives for the next application of stimulus. Record (2) exhibits the effect of moderate cooling; the transmission-time is now prolonged by ·8 second. On the assumption that the effect of cooling had remained localised, it is seen that the lowering of temperature had prolonged the time of transmission through the 10 mm. of the petiole from ·63 second to 1·43 seconds. The conductivity had thus been reduced by more than half.

In record (3) is seen the effect of a further lowering of temperature by placing small fragments of ice on the strip of cloth. The excitatory impulse initiated by the maximal stimulus of induction-shock had hitherto been unfailingly transmitted. But under the action of intense cold the impulse was arrested. In order to show that the abolition was not due to the depression of the motile excitabilty of the pulvinus, record (4) was taken of the effect of direct stimulation: the record shows that the motile excitability had undergone no change. It is thus clear that the impulse initiated by the stimulus had been arrested by the physiological depression of conductivity, induced locally by the action of cold.

Paralysis of Conductivity and Restoration by Tetanising Shocks

Experiment 40.—In connection with this subject I came across the interesting phenomenon of paralysis of conductivity as an after-effect of intense cold. After obtaining the record of the block under local application of cold, the fragments of ice were removed and the cooled portion of the petiole allowed to regain the temperature of the room, which must have been accomplished in the course of 20 minutes. After this, on taking a record of the transmitted effect of stimulation, I found that the block was still persistent. The conducting power of the benumbed tissue is thus paralysed for a period which generally lasts for about 45 minutes. I have, however, discovered the very suggestive fact that the lost conductivity can very quickly be restored by subjecting the paralysed portion of the petiole to the action of tetanising electric shocks.

ELECTROTONIC BLOCK

I have in a previous chapter described the effect of the electrotonic block in arresting conduction (*cf.* fig. 7). It was shown that by putting the electrotonic current 'on' and 'off' the transmitted excitation could alternately be arrested and allowed to proceed without hindrance. These characteristic effects were manifested in records of the mechanical response of the leaf.

I now describe a different method for demonstrating the change of conductivity effected by the electrotonic block, the record being taken on a fast-moving plate.

FIG. 27. Record of Electrotonic Block. Uppermost record normal; lowest record shows block of transmission; middle record shows restoration of conductivity on removal of block.

Experiment 41.—A maximal stimulus was applied on the petiole at a distance of 30 mm. from the pulvinus. Half-way between the point of excitation and the pulvinus were placed two polarising electrodes, 5 mm. apart, through which a constant current would be maintained constituting an electrotonic block. The first and uppermost of the three records (fig. 27) was taken without the block: the velocity of transmission was 29 mm. per second. The blocking current was next introduced. In order to prevent the excitation due to sudden make or break of the current, the blocking current, having an E.M.F. of 2 volts, was gradually introduced or withdrawn by the proper manipulation of a potential-slide. During the passage of the blocking current the testing-stimulus of induction-shock was applied, as in the first experiment of the series. The lowest record shows that there was no response, the transmission of excitation being effectively blocked. In order to show that the block persists only during the passage of the current, the latter was next gradually reduced to zero. On again repeating the stimulation the block was found to be no longer operative (middle record) and the excitation was transmitted at practically the same rate as at the beginning. The slight increase in the rate is probably due to facilitation caused by previous stimulation.

Block by Local Application of Poison

Reference has been made to the inconclusive character of Pfeffer's narcotisation experiment. The ineffectiveness of the block, it was explained, might have been due to the thickness of the petiole preventing free access of the anæsthetic to the conducting elements in the interior. It occurred to me that the physiological block induced by a drug could be made more effective by the employment of solutions of strongly toxic agents, like copper sulphate or potassium cyanide. The choice of a strong poison was deemed advisable, because the absorption of even a small quantity might then prove effective in inducing a depression culminating in abolition of conduction. The application of chloroform has the drawback that its vapour is always escaping, with resultant stoppage of absorption; an additional drawback is that the spreading vapour renders the motile organ insensitive. There is no such disadvantage in the employment of a non-volatile poison like copper sulphate or potassium cyanide in solution, which by local application would affect the conductivity of the petiole without modifying the motile sensibility of the pulvinus. The poisonous solution was applied on a strip of cloth 10 mm. wide, wrapped round the petiole, midway between point of stimulation and the pulvinus.

I will now describe experiments on the effect of poison applied on a narrow zone of the petiole of Mimosa.

Successive records of transmission-time were taken at intervals of 20 minutes, before and after subjecting an intermediate portion of the petiole to the action of poison. Effective stimulus of induction-shock of intensity of 2 units was applied on the petiole, generally at a distance of 30 mm. from the pulvinus. The first record of the series gives the velocity of normal conduction; the second and the subsequent records exhibit the progressive action of the toxic agent.

Experiment 42. *Copper sulphate solution.*—The normal record (1) in fig. 28 shows response to have taken place

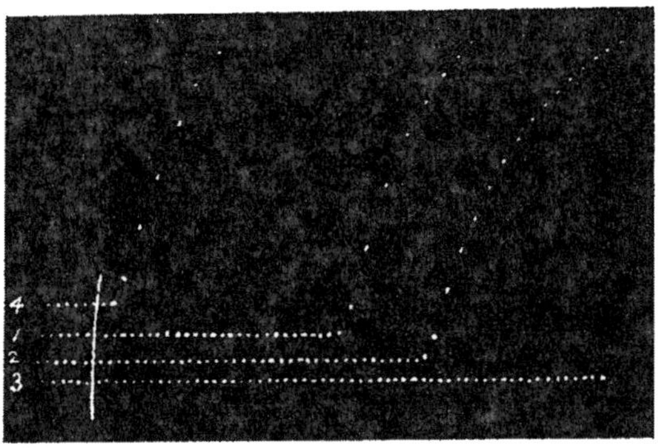

FIG. 28. Effect of Copper Sulphate solution in the retardation and final arrest of Conduction.

(1) Normal record; (2) Retardation caused by 20 minutes' application; (3) Arrest caused by application for 40 minutes; (4) Record of direct stimulation.

27 spaces after the application of the stimulus, the interval between the successive dots being ·1 second. The total time was therefore 2·7 seconds. Subtracting from this the latent period ·15 second, then 2·5 seconds is the actual transmission-time through 30 mm. The time of transmission through 10 mm. is therefore ·83 second.

Record (2) of the series shows the effect of application of copper sulphate solution for 20 minutes on a portion of the petiole 10 mm. in breadth. It will be noticed that the transmission-time was prolonged by 10 spaces, *i.e.* by 1 second. Assuming that the effect of the poison was localised, it appears that it had, during 20 minutes' application, increased the time of transmission through 10 mm. from ·83 to 1·83 seconds. By the absorption of a small quantity of poison the conductivity had thus been reduced by more than 50 per cent.

Record (3) of the series was taken after a further period of 20 minutes. The transmitted excitation is seen to be completely blocked, as the result of the application of copper sulphate for 40 minutes. In order to show that the absence of response was due not to the abolition of motile excit-

FIG. 29. Abolition of Conductivity by the action of Potassium Cyanide.

(1) Normal record; (2) Arrest of conduction after application for five minutes; (3) Record showing arrest of impulse, even with strong stimulus; (4) Record of direct stimulation.

ability of the pulvinus, but to the block of conductivity of the petiole, a fourth record was taken under direct stimulation, which proves that the motile excitability of the leaf had remained unimpaired.

Experiment 43. *Mercuric chloride solution.*—A series of records was taken exhibiting the effect of mercuric chloride solution. Conduction was found to be arrested after a period of application so short as 10 minutes. The record obtained was very similar to that given in fig. 29.

Experiment 44. *Potassium cyanide solution.*—I also took a series of records exhibiting the effect of strong solution of potassium cyanide. Record (1) gives the normal transmission-time (fig. 29). Record (2) was taken, as before, after allowing 20 minutes for recovery. The poisonous solution was, however, applied 15 minutes after the previous record; consequently the second record shows the effect of application of potassium cyanide for a period of 5 minutes only. The effect of this poison on the conductivity of the petiole of Mimosa was so great that even with such a short application, the transmission of excitation caused by maximal stimulus of two units was completely blocked. Record (3) was taken with the secondary coil pushed close to the primary, the stimulus-intensity being thus raised to 15 units. Even under this intense stimulation conduction was found to be arrested. Record (4) was obtained under direct stimulation. The response shows that the motility of the pulvinus had undergone no change. It is thus clear that the abolition of response to indirect stimulation was solely due to the abolition of conductivity induced by the action of the poison.

SUMMARY

Rise of temperature, within limits, increases the velocity of transmission of excitation in Mimosa; conversely, lowering the temperature diminishes the velocity.

The transmission of excitation can be arrested by inducing physiological block in the conducting tissue.

Under excessive lowering of temperature, the power of conduction becomes paralysed; the paralysis persists for a time even after return to a normal temperature. The condition of paralysis can be quickly removed by tetanising induction-shocks.

Conduction is inhibited by the interposition of an electrotonic block. On the stoppage of the blocking current the normal velocity of conduction is restored.

Application of poison, at an intermediate region in the path of conduction, at first lowers the velocity: the depression culminates in permanent abolition of conduction in the poisoned area.

The time required for abolition of conduction depends on the virulence of the poison employed. The abolition is much more rapid under potassium cyanide than under copper sulphate solution.

CHAPTER VIII

POSITIVE AND NEGATIVE IMPULSES UNDER INDIRECT STIMULATION

WHEN a stimulus is applied directly on the pulvinus of Mimosa, the leaf undergoes an almost instantaneous fall. The latent period, *i.e.* the interval between the application of stimulus and the resulting response, is about 0·1 second. Indirect stimulation, *i.e.* application of stimulus at a distance from the pulvinus, also causes a fall of the leaf; but a longer interval elapses between the incidence of stimulus and the response, for a certain length of time is required for the transmission of the generated impulse through the intervening conducting tissue.

Conductivity varies widely in different plants and their organs. In thin petioles of Mimosa the velocity has been shown to be as high as 400 mm. per second. In the stem the velocity is considerably less, about 5 mm. per second in the longitudinal direction; but conduction across the stem is a very much slower process. I have shown that a stronger or a more prolonged stimulation is necessary to overcome the resistance offered to the passage of excitation in a transverse direction (Experiment 18). The velocity of conduction in other plants is very much lower. In the petiole of Averrhoa the longitudinal velocity is of the order of 1 mm. per second.

MECHANICAL RESPONSE UNDER VARIATION OF TURGOR

The mechanism of movement of the leaf depends ultimately upon variation of turgor in the pulvinus. The following experiments demonstrate the characteristic movements produced by changes of turgor of the motile organ.

Experiment 45. *Effect of increase of turgor.*—A young Mimosa plant was carefully removed from the ground and placed, with the soil adhering to the roots, in a linen bag. This was held securely by a clamp, and one of the leaves attached to the recorder. Water was withheld for a day,

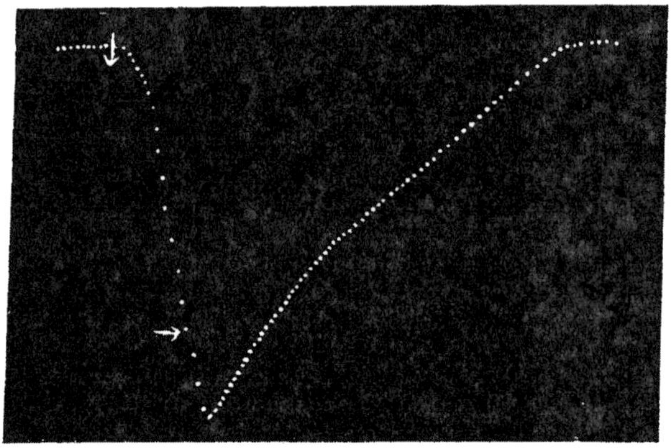

FIG. 30. Response of pulvinus of *Mimosa pudica* to variation of turgor.

Increased turgor by application of water at point marked with vertical arrow induced erectile movement. Diminution of turgor by application of KNO_3 solution at the point marked with the horizontal arrow, brought about the fall of the leaf within 80 seconds. Successive dots at intervals of 5 seconds. (The down-curve represents up-movement and *vice versâ*.)

the result being a general loss of turgor throughout the plant. A vessel filled with water was now raised from below so as to moisten the roots. The absorption of water led to an increased turgor of the pulvinus, which caused an *erectile* movement of the leaf. The distance between the immersed portion of the plant and the leaf was 20 mm., and the up-movement of the leaf was initiated 10 seconds after the application of water.

Experiment 46. *Effect of diminution of turgor.*—While the leaf was in the process of erection, a change was produced by substituting KNO_3 solution for the water supplied to the roots. The plasmolytic withdrawal of water gave rise to a wave of diminished turgor, the effect of which became perceptible within 40 seconds, by the responsive *fall* of the leaf (fig. 30).

Thus the leaf is capable of two distinct mechanical reactions: (1) an erectile or positive movement indicating expansion and enhancement of turgor, and (2) a fall or negative movement indicating contraction and diminution of turgor. The motile leaf or leaflet is merely an indicator which manifests the induced changes of turgor in the pulvinus.

Dual Character of the Transmitted Impulse

Direct stimulation induces a sudden diminution of turgor resulting in a sudden fall of the leaf. A strong indirect stimulus applied to the highly conducting petiole of Mimosa gives rise to an impulse which quickly reaches the pulvinus and, like direct stimulation, causes a fall of the leaf. This excitatory impulse may, since it induces a negative response, be distinguished as *negative*. When, however, a moderate stimulus is applied to a less actively conducting tissue, such as the semi-conducting petiole of Averrhoa, and at a considerable distance from the leaflets, the impulse transmitted from the point of stimulation causes an erectile movement of the leaflets as it reaches them; that is to say, the response is not negative but *positive*, and the impulse inducing it may be distinguished as *positive*. But when the distance of transmission is somewhat reduced, the response is not single but double; the erectile, positive movement is followed by a fall or negative movement.

I here give a detailed account of a typical experiment carried out with *Averrhoa Carambola*, which clearly brings out the characteristic effects of Indirect Stimulation.

Experiment 47.—Stimulus of electric shock applied at a point on the long petiole of Averrhoa causes the successive fall of pairs of leaflets. In the experiment to be described, one of the leaflets of the plant was attached to a recorder which gave a magnification of 30 times. Stimulus of electric shock was applied at a distance of 50 mm. from the leaflet. The successive dots in the record are at intervals

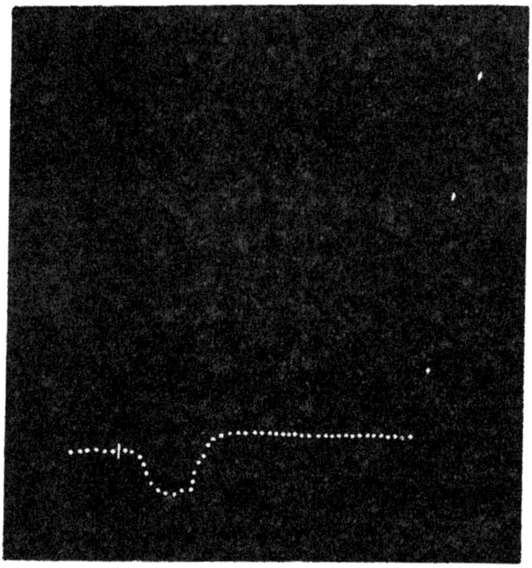

FIG. 31. Dual Response under Indirect Stimulation.

Postive erectile response represented as a down-curve: the more intense negative response of fall shown as an up-curve. (Averrhoa.) Successive dots at intervals of one second. Stimulation indicated by short vertical line.

of 1 second (fig. 31). The record shows that an identical stimulus causes two responses of the distant leaflet, the positive followed after a considerable interval by the negative. The positive or erectile response occurred 1·5 seconds after stimulation, the recovery being completed in 16 seconds. After a further delay of 26 seconds, during which the responding leaflet remained quiescent, the normal negative response occurred by the excitatory fall of the leaflet. These characteristic results lead to the conclusion that indirect stimulation gives rise to two impulses and that they travel with different velocities. The movement in response to the positive impulse is slow, whereas that due to the negative is very abrupt, almost 'explosive,' the successive dots in the record being relatively far apart. As regards the velocity of transmission the relation is reversed, the positive being the quicker of the two. In the present case, the velocity of the excitatory *negative* impulse was 1·1 mm. per second, as against 33 mm. for the *positive* impulse.

Effect of Distance on Transmission

In the last experiment the stimulus was applied at the moderate distance of 50 mm. When it was applied at a greater distance the response was positive only, for the reason that, in a tissue whose conducting power is not great, the excitatory (negative) impulse is weakened even to extinction in transmission through a long distance, whereas the positive impulse is not weakened to the same extent by long transmission. The negative impulse may thus fail to reach the responding organ, whilst the positive does so, in which case stimulus applied at a distance will give rise only to a positive response. When, however, the distance of transmission is reduced, a different result is obtained. The negative now overtakes the positive impulse, and, since it is the more intense of the two, the feeble positive response will be masked by the superposed negative. The separate exhibition of the two responses is therefore only possible when there is a sufficient lag of the negative impulse behind the positive. The lag increases with the increase of distance of transmission and decreases with its diminution. Application of stimulus near the responding organ therefore gives rise only to a negative response, which masks the positive.

From what has been said it will be understood that the exhibition of positive response is favoured when the transmitting tissue is semi-conducting and the stimulus feeble. It is thus easier to exhibit the positive effect with the feebly conducting petiole of Averrhoa than with the better conducting petiole of Mimosa. It is, however, possible to obtain positive response in Mimosa under suitable conditions. The positive response can be obtained with the stem, in either longitudinal or transverse conduction. The rate of longitudinal transmission in the stem is much slower than it is in the petiole, hence under sufficiently feeble stimulus only a positive response is obtained; under stronger stimulus it is followed by a negative response. It should be remembered that previous stimulation enhances conductivity, so that ineffective conduction becomes gradually effective; the response may thus become diphasic, that is, positive followed by negative (Experiment 48).[1]

The transverse conductivity of the stem is, as has been shown, very feeble. Hence positive and diphasic responses can be easily obtained when the conduction is in a transverse direction (Experiment 49).

Positive Impulse in Stem of Mimosa: Longitudinal Transmission

Experiment 48.—The testing stimulus was applied to the stem at a distance of 10 mm. below the responding leaf. The intensity of stimulus in the three successive experiments was increased from 1 to 5 and then to 8 units. The first and the lowest record of the series (fig. 32) was taken under the stimulus-intensity of 1. Under this relatively feeble stimulus, a positive or erectile response, indicative of increase of turgor in the pulvinus, was alone induced, 0·7 second after stimulation; there was no indication whatsoever of the occurrence of a negative response. After the usual interval of 20 minutes the next or middle record was taken under a stronger stimulus of 5 units. The

[1] Cf. *Plant Response*, p. 535; *Comparative Electro-Physiology*, p. 64: *Irritability of Plants*, p. 196.

response was now diphasic, positive followed by negative; the positive took place 0·6 second, and the negative 9·4 seconds, after the application of stimulus. Finally, when the stimulus-intensity was raised to 8 units, the positive

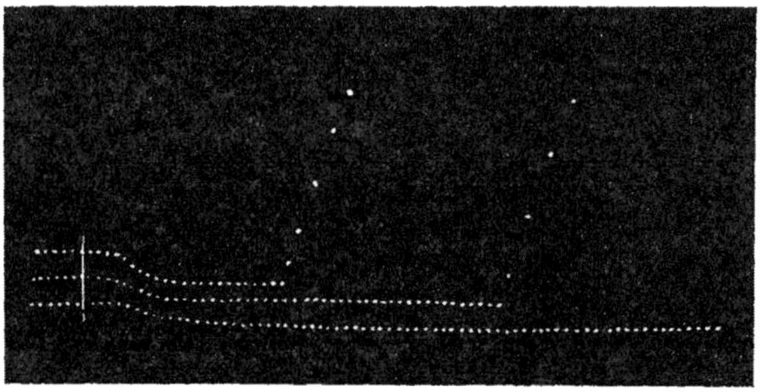

FIG. 32. Effect of Intensity on Response (Mimosa).
Lowest record under stimulus-intensity of 1, response positive; middle record under 5, response diphasic; uppermost record under 8 units, response diphasic with shortened interval. Vibration-frequency 5 times per second.

response took place after the same interval as before, but the negative or excitatory response occurred earlier than in the last case, that is to say, after an interval of 4·6 seconds instead of 9·4 seconds.

POSITIVE IMPULSE IN THE PETIOLE OF MIMOSA

Experiment 49.—I have obtained a positive response to transmitted impulse even in the highly conducting petiole. The stimulus in this case has to be considerably reduced. The difficulty in the detection of the positive response arises from its small amplitude. This drawback is completely removed by the employment of my Optical Lever ('Plant Response,' p. 5), which magnifies the responsive movement of the leaf more than a thousand times. The optical lever affords many advantages for demonstration; the weight of the lever is slight, and there is practically no friction. When a sub-minimal stimulus is applied to a sub-petiole of the leaf, the transmitted impulse is found to give rise to an erectile response of the leaf, shown by the up-movement of the reflected spot of light through about 15 cm. A stronger stimulus causes a brief positive followed by a very rapid and large negative response.

An interesting result is obtained by successive applications of the sub-minimal stimulus. The first responses are positive, but on account of Facilitation induced by stimulation, the conductivity is gradually increased. Hence, under successive applications of sub-minimal stimulus, the response becomes gradually transformed from positive to negative through the intermediate diphasic.

The positive impulse in the petiole can be more easily demonstrated when the specimen is in a sub-tonic condition than when it is normal.

Positive Impulse in Stem of Mimosa: Transverse Transmission

Experiment 50.—A thin stem of Mimosa was held vertical by means of a clamp. A narrow beam of light from a small arc-lamp was made to fall upon it at a point exactly opposite to the motile leaf which was to serve as the indicator of response to positive or negative impulses transmitted across the stem. The leaf was attached to the recording lever, the successive dots in the record being at intervals of 10 seconds. Stimulation by light caused a positive or erectile movement in about 10 seconds after the application of the stimulus. When the stimulus was moderate or of short duration, the response remained positive. But with strong or prolonged stimulation the excitatory negative impulse was conducted across the stem to the distal side, causing a very rapid fall of the leaf (fig. 33). In the present case the negative impulse reached the motile indicator in the course of about 3 minutes. The transmission is more rapid under strong stimulus. Similar effects were also obtained under unilateral electric or thermal stimulation.

In order to obtain the preliminary positive response, care has to be taken that the stimulus is of moderate intensity. In this respect, photic stimulus offers special advantages, since its intensity can be so adjusted as not to be excessive.

The experiment that has just been described is of much significance. Though the stem of Mimosa exhibits no

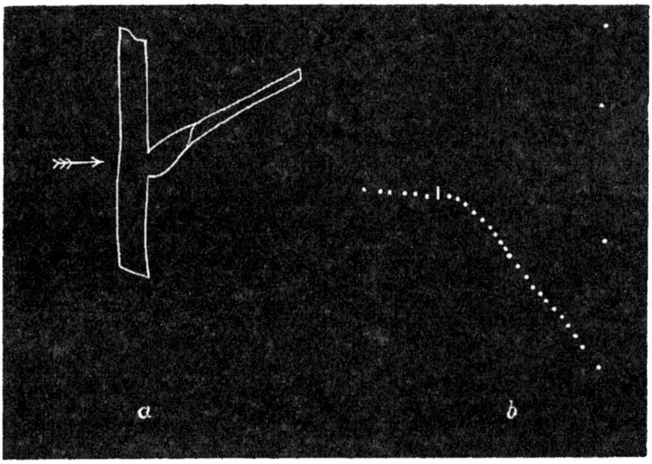

Fig. 33. Erection of Mimosa Leaf due to Indirect Stimulation.
(a) Diagram of the experiment: point of application of stimulus indicated by arrow; (b) Erectile response (shown by down-curve) followed by rapid fall (up-curve). Successive dots at intervals of 10 seconds.

movement, and so may appear to be insensitive to stimulation, yet its perception and its reaction to stimulus are shown by the fact that, in response, it can generate and transmit two characteristic impulses: one of these is positive, and gives rise to an enhancement of turgor at the distal side; the other, negative, induces the opposite reaction of diminution of turgor. These characteristic reactions to unilateral stimulation will be shown to offer the most satisfactory explanation of tropic curvature.

Effect of Direct and Indirect Stimulation on Growth

I have shown elsewhere [1] that increase of turgor, within limits, enhances the rate of growth, while diminution of turgor lowers the rate. I have also shown that direct stimulation induces a retardation of growth which may culminate in actual contraction. Since indirect stimulation gives rise to an enhancement of turgor, it induces an acceleration of the rate of growth.

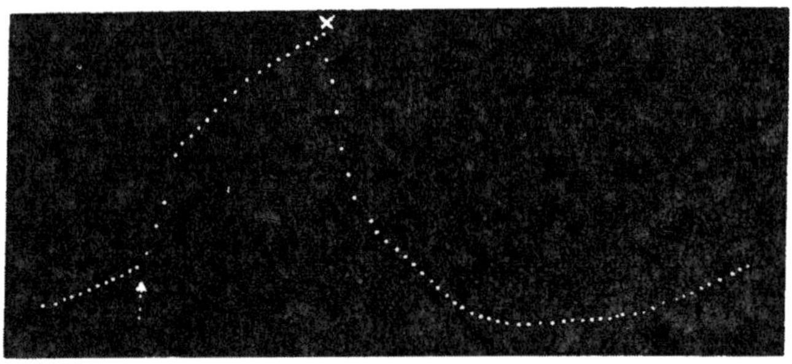

FIG. 34. Effects of Direct and Indirect Stimulation on Rate of Growth.

Normal rate of growth shown in first part of record; enhancement of the rate, due to indirect stimulation at arrow, seen in sudden erection of the curve. Direct stimulation at cross induced a sudden contraction shown by the down-curve.

Experiment 51.—The above facts are demonstrated in a record (fig. 34) obtained with my High Magnification Crescograph. Indirect stimulation enhanced the rate of growth, as shown by the resulting erect curve. Direct stimulation induced, on the other hand, an actual contraction, as seen in the reversal from an ascending to a descending curve. When the indirect stimulation is sufficiently strong or prolonged, the response is diphasic, a preliminary acceleration followed by a depression of growth culminating in an actual contraction.

[1] *Life Movements in Plants*, vol. i. (1918), p. 191.

Heliotropic Curvature under Unilateral Stimulation

It has just been shown that while a growing organ shows retardation of growth and contraction when directly stimulated, under indirect stimulation it exhibits acceleration of growth and expansion. Now a growing organ under unilateral stimulation, say, by light, is stimulated directly at the proximal and indirectly at the distal side: the rate of growth of the former is retarded, whilst that of the latter is accelerated. The result is a positive heliotropic curvature. Under intense or long continued stimulation, the excitation reaches the distal side, neutralising the positive curvature, or reversing it to negative.

Response of the Pulvinus of Mimosa to Unilateral Stimulation

Broadly speaking, three types of tissue may be physiologically discriminated in plants. There is, first, the indifferent parenchyma in which, on account of the numerous septa, excitation remains more or less localised at the point of application of stimulus. At the other extreme, there is the highly conducting nervous tissue, in which excitation is transmitted to a considerable distance, the conducting nerve exhibiting no visible change during the passage of the impulse. Between these two extremes there is an intermediate type, such as the contractile cells of the pulvinus, in which the wave of excitatory contraction passes from cell to cell, at a rate slower than that of the nervous impulse. I may, for convenience, distinguish this as *cellular propagation of excitation*. The phenomenon is not unlike the propagation of a wave of contraction from cell to cell in the muscle of the animal heart.

The pulvinar tissue may be regarded as semi-conducting, and I will now describe its characteristic response to indirect stimulation applied locally on the upper surface of the pulvinus. Since the upper half is relatively inexcitable, compared with the lower half, for simplicity of explanation I confine attention to the reaction of the more excitable half of the organ.

Experiment 52.—A narrow beam of strong light from a small arc-lamp was thrown on the upper half of the

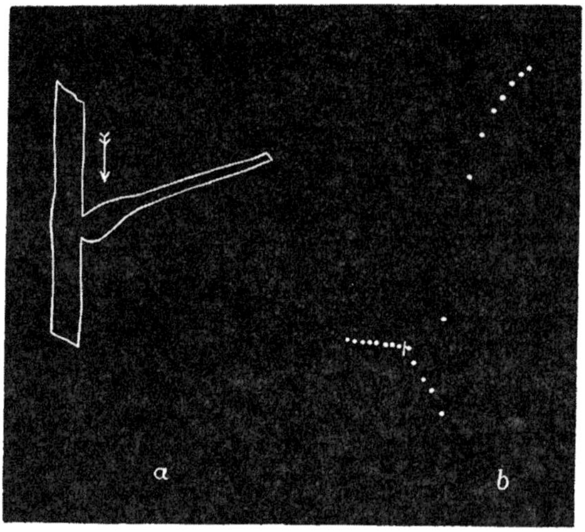

FIG. 35.—Record of effect of continuous application of light on upper half of pulvinus of Mimosa.

Note positive erectile response (down-curve) followed by neutralisation and pronounced reversal into negative due to transverse conduction of excitation.

pulvinus, and record of the response was obtained in the usual manner. The record (fig. 35) shows that after a latent period of 5 seconds, a positive erectile response was initiated, due to the expansion of the lower half of the pulvinus. Under the continued action of light, the excitatory impulse reached the more excitable lower half and caused very rapid negative response shown by the fall of the leaf, due to contraction of that half of the organ. In sensitive specimens the negative response is so abrupt and intense, that the writing lever is jerked off beyond the recording plate before making a dot on it. The thickness of the pulvinus was 1·5 mm., and the distance which the excitatory impulse had to traverse to reach the lower half was therefore about 0·75 mm. The time for the transverse transmission of excitation under the action of light was found to vary in different cases from 50 to 80 seconds. The velocity of the transverse transmission of excitation in the pulvinus is therefore of the order of 0·011 mm. per second, which is very much lower than the velocity in the petiole of the plant.

Other modes of stimulation gave similar results. Stimulation without mechanical disturbance can be produced by the application of a drop of dilute solution of hydrochloric acid on the upper half of the pulvinus. This produces a preliminary positive followed by a more intense negative response. The rate of transverse transmission becomes considerably enhanced under the action of a strong stimulus.

SUMMARY

By the method of mechanical response it has been shown that, in plants having motile leaves, a single stimulation gives rise to two impulses, positive and negative. The positive travels at a quicker rate and induces an enhancement of turgor and erectile response of the leaf. The negative or excitatory impulse travels at a slower rate and induces a diminution of turgor and a rapid fall of the leaf. The negative response is far more intense than the positive.

A feeble or sub-minimal stimulus gives rise only to the positive response. A minimal stimulus gives rise to a diphasic response, positive followed by negative.

When the distance of transmission is reduced, the predominant negative impulse overtakes and masks the feeble positive.

Direct stimulation, inducing a diminution of turgor,

causes a retardation of the rate of growth. Indirect stimulation of feeble intensity causes an acceleration.

The above facts explain tropic curvatures produced by unilateral stimulation; under feeble or moderate stimulation the curvature is positive; under strong and long continued stimulation the positive curvature passes over into negative.

Local stimulation of the upper half of the pulvinus of Mimosa gives rise to a positive response. Under strong or long continued stimulation the positive is converted into an intense negative response in consequence of the conduction of excitation to the more excitable lower half of the organ.

The Laws of Direct and Indirect Stimulation are as follows:

(1) THE EFFECT OF ALL FORMS OF DIRECT STIMULATION IS A DIMINUTION OF TURGOR, CONTRACTION, DIMINISHED RATE OF GROWTH, AND THE NEGATIVE MECHANICAL RESPONSE OF THE FALL OF THE LEAF.

(2) THE EFFECT OF INDIRECT STIMULATION OF FEEBLE INTENSITY IS AN INCREASE OF TURGOR, EXPANSION, ACCELERATED RATE OF GROWTH, AND THE POSITIVE MECHANICAL RESPONSE OF ERECTION OF THE LEAF.

(3) PROLONGED APPLICATION OF STIMULUS OF MODERATE INTENSITY GIVES RISE TO A DIPHASIC RESPONSE, POSITIVE FOLLOWED BY NEGATIVE.

(4) IF THE INTERVENING TISSUE BE HIGHLY CONDUCTING, THE EFFECT OF THE TRANSMITTED POSITIVE IMPULSE BECOMES MASKED BY THAT OF THE PREDOMINANT NEGATIVE.

CHAPTER IX

ELECTRIC RESPONSE OF ORDINARY PLANTS TO DIRECT AND INDIRECT STIMULATION

THE mechanical response of 'sensitive' plants to direct and indirect stimulation has been fully described in the last chapter. It was shown that under direct stimulation the irritable parenchyma of the pulvinus exhibits a diminution of turgor, a contraction, and a negative mechanical response exhibited by the fall of the leaf. Under indirect stimulation the response was either positive, diphasic, or negative, according to the intensity of stimulus, the conductivity of the intervening tissue, and the tonic condition of the plant.

These results were demonstrated in sensitive plants, in which the movement of the motile organ gives a conspicuous indication of the character of the transmitted impulse. The question now arises: are these specific reactions manifested only by sensitive plants, or are they to be found universally in all plants?

ELECTRIC RESPONSE TO DIRECT STIMULATION

It used to be thought that 'sensitive' plants alone were excitable, their motor organs exhibiting excitatory reaction by contraction and by concomitant electric change of galvanometric negativity. Ordinary plants, on the other hand, were regarded as irresponsive, both mechanically and electrically.

I have, however, been able to show [1] that all plants and all their organs are excitable, the state of excitation being manifested by an electric response of galvanometric negativity. To obtain an indubitable demonstration of electric response a non-electrical mode of stimulation is essential;

[1] *Response in the Living and Non-Living*, 1902.

otherwise the leakage of the stimulating induction-current might produce a galvanometric deflection not due to true excitation. This difficulty was overcome by a device by which the plant could be subjected to mechanical stimulation, the intensity of which could be maintained uniform or increased in a graduated manner.

Experiment 53.—I obtained uniform electric responses from the petiole, stem, roots and fruits of various plants

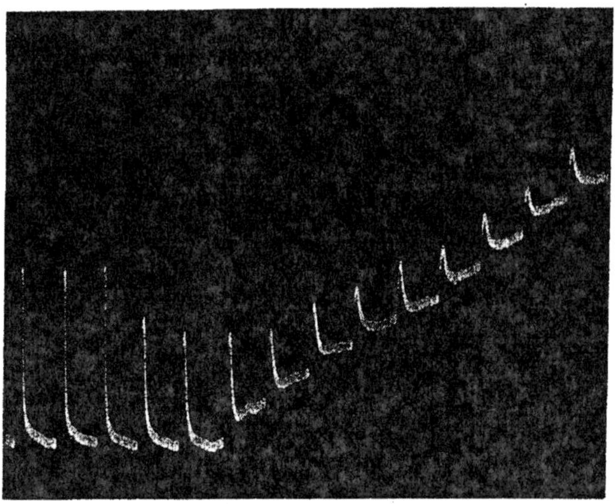

FIG. 36. Galvanometric Record of effect of Chloroform on Electric Response of Carrot.
First three responses are normal; subsequent depression produced by chloroform.

under uniform stimulation. In order to prove that the electric response of galvanometric negativity is physiological, it was ascertained that depression was induced by the action of chloroform. The first three responses (fig. 36) were normal; the application of chloroform produced rapid diminution in the amplitude of response, which completely disappeared on the death of the tissue.

Having obtained the response to direct stimulation in ordinary plants, the next point is whether or not induced excitation is transmitted to a distance; and, if so, what are its characteristics. By means of the mechanical response given by sensitive plants with motile leaves, it was ascertained that two distinct impulses, a positive and a negative, were generated under indirect stimulation, and could be discriminated under suitable conditions. What has now to be determined is whether or not this evidence is corroborated by the results of the investigation of the transmitted impulse by the electrical method. Should the various modifications of the transmitted impulse, manifested in the mechanical response, find their parallel in the electrical response, then, clearly, both forms of response must have a common source.

Positive Response in Semi-conducting or Non-conducting Tissues

Experiment 54.—The indifferent tissue of the lamina of the leaf is practically a non-conductor. Two electric

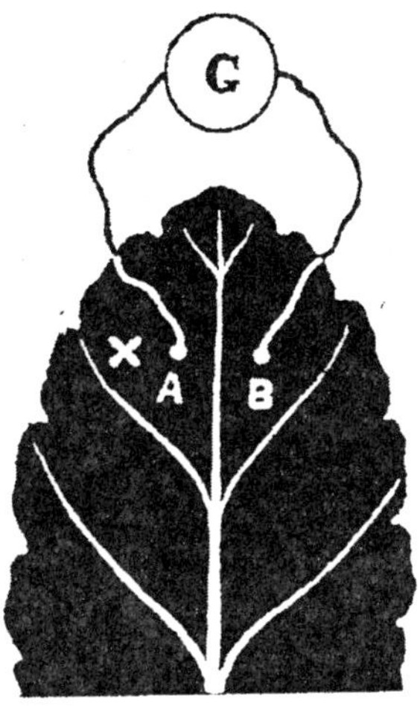

FIG. 37.

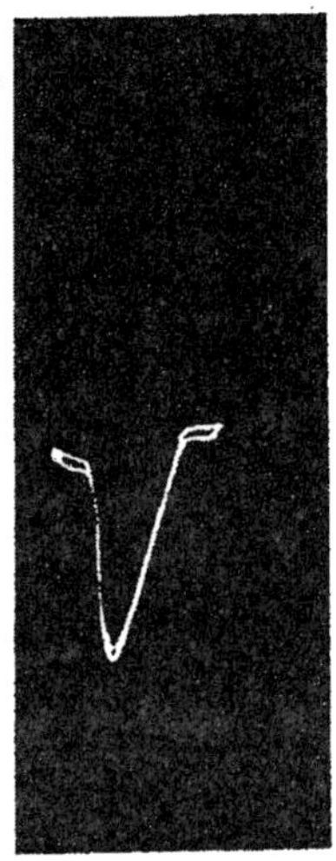

FIG. 38.

FIG. 37.—Electric connections for response of non-conducting tissue. Stimulus applied at cross near A.

FIG. 38.—Positive Electric Response of non-conducting tissue (down-curve).

contacts are made at two points A and B on the lamina (fig. 37); a highly sensitive reflecting galvanometer is then put in circuit, the sensitiveness of which is such that a current of 10^{-10} ampere gives deflection of 1 mm. at a distance of 1 metre. The heated tip of a glass rod is momentarily applied at a distance of 5 to 10 mm. from the contact A. The response (down-curve) is the expression of an electric change of galvanometric positivity. Recovery takes place after the passage of the impulse (fig. 38).

I will now describe experiments on the impulse transmitted along highly conducting tissue.

Diphasic and Monophasic Response

The excitatory impulse is propagated from point to point in an animal nerve, each excited point becoming galvano-

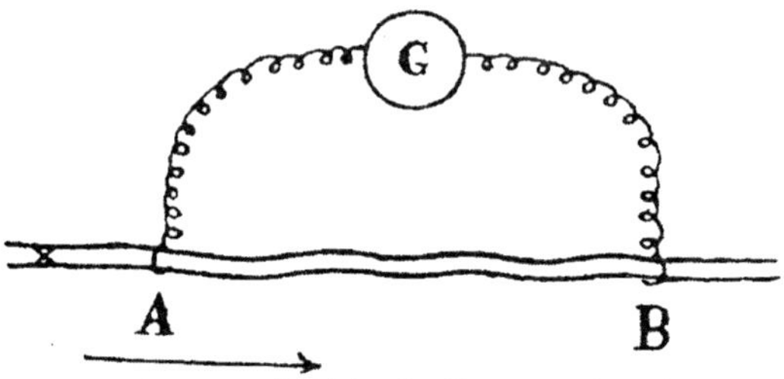

FIG. 39.—Diagram of Diphasic Response in animal nerve.

metrically negative to any point at rest. The wave of excitation is thus accompanied by a wave of galvanometric negativity. The result is shown diagrammatically in fig. 39, in which stimulus is applied at a point marked with a cross near the electrode A. The wave reaches A the earlier, and the deflection of the galvanometer shows that A is negative compared with the unexcited point B. The wave soon moves past A and reaches B, making that point negative in respect of A; this causes a reversal of the former galvanometric deflection. If the conducting power of the nerve be very high, the two opposite impulses act on the galvanometer in quick succession; and since the galvanometric inertia is considerable, the resultant deflection is practically zero.

This difficulty in obtaining a record of electric response to transmitted impulse may be overcome in three different ways:

First, by the employment of an electric recorder in which the inertia of the indicator is very slight, as in the string-galvanometer of Einthoven. The diphasic response, namely the negative response at A followed by that at B, may then be obtained on a moving photographic plate. The negative response at A then appears, say, as an up-curve, and that at B, as a down-curve.

Secondly, a monophasic response at A, unopposed by any at B, may be secured by abolishing the excitability of B, as by causing local injury. This method introduces several complications: a current of injury is produced, and the response at A is then observed as a negative variation of the current of injury. The injured tissue may, however, exhibit gradual recovery, on account of which the resultant response undergoes diminution or abolition. A fresh injury has therefore to be inflicted at or near B, to abolish excitability at that point, and thus obtain the unopposed response at A.

Thirdly, these drawbacks can be overcome by employing the Method of Natural Block, in which a non-conducting tissue is interposed between the first contact A and the second contact B. The excitatory impulse is then unable to reach the second contact. The specimen, in such a case, is uninjured, and the response of A is perfectly normal.

It is necessary to find suitable means of stimulation to initiate the transmitted impulse. For merely qualitative purposes, stimulation may be effected by the application of a heated rod, or by the application of a drop of acid; but these, as already stated, are incapable of repetition with the same intensity.

Stimulation by Thermal Shock

In the attempt to define the fundamental electric reactions to transmitted excitation, it is desirable that the stimulus should be non-electrical and at the same time capable of graduation in intensity. I have succeeded in perfecting a method of stimulation by thermal shock. A single loop of thin platinum wire may be made closely to surround the experimental tissue: or a V-shaped platinum wire may be applied on the conducting tissue, a film of water being interposed between the platinum wire and the tissue to prevent scorching. A definite electric current sent through the platinum loop for a given length of time gives rise to a sudden thermal variation which acts as a stimulus. This method is most convenient when successive stimulations of equal intensity are required, the *duration* of the heating current being regulated by a metronome. The *intensity* of the stimulus is graduated in a predetermined manner by the adjustment of the heating current, taking care that the platinum wire is raised in temperature only through a few degrees, so as not to injure the tissue. The required temperature can be estimated, in practice, by touching the platinum with one's finger, a heat that can be felt without inconvenience. Excitation is produced in this way either by one, or by a summated series of thermal shocks.

Having indicated the most suitable method of obtaining electric response and having eliminated all complications by the employment of a non-electric mode of stimulation, the following experiments were made which show that the conducting tissue of the plant exhibits all the responsive characteristics of the animal nerve.

Electric Response of Petiole-laminar Preparation

The typical reactions are best demonstrated with an ordinary leaf, in which the parenchymatous tissue of the lamina is practically a non-conductor. The midrib is, on the other hand, a conductor of excitation: in it the conducting phloem surrounds the xylem, the bundle being concentric. The electric connections are shown in fig. 40.

In (a) the method of natural block is employed to obtain monophasic response, that is, one electrode is placed on the midrib, the other on the lamina; in (b) both the

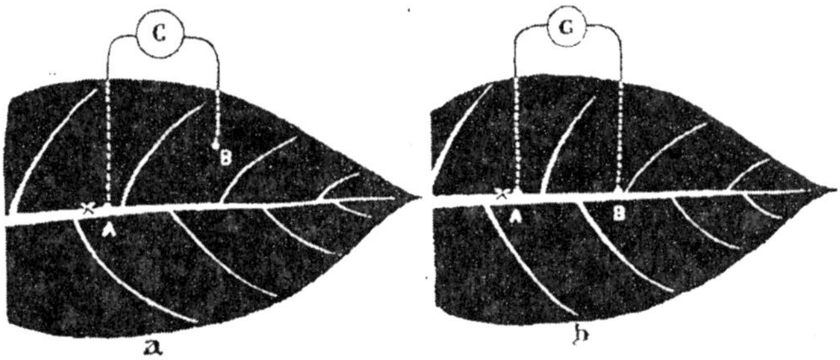

FIG. 40.—Electric Connections on Leaf for (a) Monophasic and (b) Diphasic response. Stimulus applied at cross.

contacts A and B are made on the conducting midrib to obtain the diphasic response.

Experiment 55. *Monophasic response.*—The record (fig. 41) gives successive responses to uniform thermal

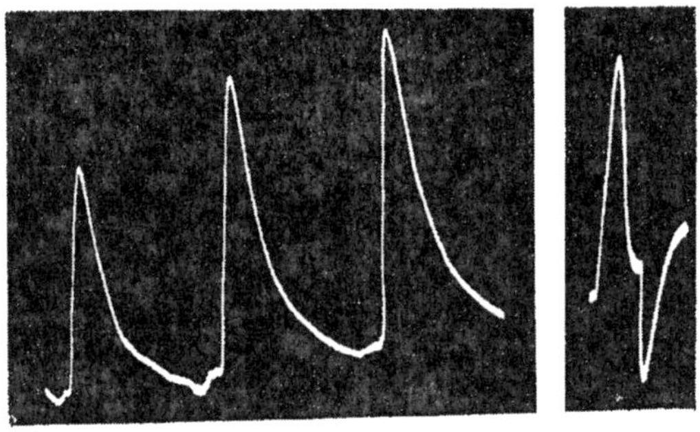

FIG. 41.—Electric Response of Midrib to indirect stimulation. Left record shows monophasic response exhibiting staircase-increase. Right record exhibits diphasic response.

stimuli, showing a staircase-increase due to Facilitation as the after-effect of stimulation. In experiments on indirect stimulation it is generally found that the conduction is ineffective in the beginning, the response being positive. The resistance or block is gradually removed by repeated stimulation, generally in four stages. In the first stage the response is positive; in the second it is diphasic; in the third it is negative, exhibiting a staircase-increase; and in the final or fourth stage, when the conductivity attains an optimum value, the responses become uniform and maximum.

Experiment 56. *Preferential conductivity.*—A very interesting result is obtained by the application of stimulus at the mid-point between A and B (fig. 40, b). It is found that while the centrifugal excitatory impulse of a moderate stimulus reaches B, there is no such transmission in a centripetal direction towards A, which remains unaffected. It is only under stimulation of stronger intensity that conduction takes place in both directions.

Experiment 57. *Diphasic response.*—When the two contacts A and B are both made on the midrib (fig. 40, b), the excitation first reaches A, producing negativity at A with a resulting up-curve in response: the excitation then reaches B, and a down-curve is obtained, indicative of the later negativity at B. A record of the diphasic response is given in fig. 41. The point B should be at a sufficient distance from A, otherwise the two impulses on the galvanometer would neutralise each other.

Experiment 58. *Time-relations of the Monophasic response.*—The wave of excitation passes from point to point along the conducting tissue; the excitation of any point is therefore transitory. The matter of interest is to find out the rate at which it reaches its maximum, and the rate at which it declines. It should be remembered that the excitatory process in the plant is relatively slower than in the highly conducting animal nerve. Again in the midrib of an ordinary leaf, conduction is, generally speaking, slower than in the petiole of Mimosa. The rise and fall of excitation at any point can be determined by the concomitant electric variation. The inertia of the moving indicator of an ordinary galvanometer is too great for this determination. The string-galvanometer, on the other hand, offers special advantages in such investigation, since its response is practically instantaneous. The only drawback is that the sensitivity of the apparatus is very much lower than that of a highly sensitive galvanometer. The electromotive variation in some plants is, however, sufficiently high to be recorded even by the Einthoven galvanometer.

The monophasic record of the midrib obtained with the Einthoven galvanometer is reproduced in fig. 42. The time-marks below indicate intervals of tenths of a second. The maximum excitation was attained, in this case, in the course of 0·8 second, the recovery being completed in the further period of 4·3 seconds.

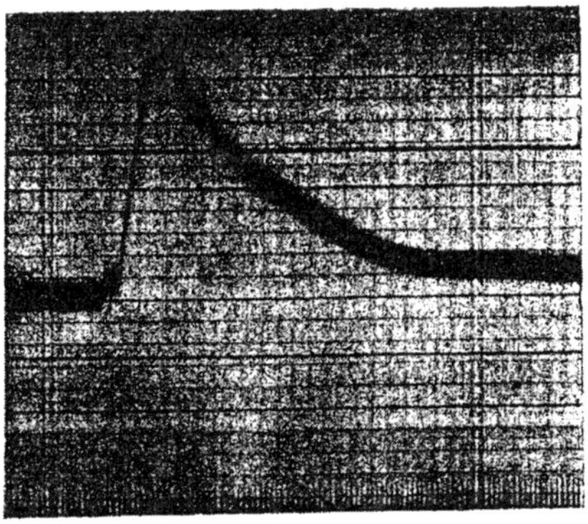

FIG. 42.—Time-relations of Monophasic Response of Midrib obtained by Einthoven String-galvanometer.

Time-marks, a tenth of a second.

The results here given prove that the characteristics of transmitted impulse, as ascertained from the mechanical response of motile sensitive plants, find an exact parallel in the electric response of ordinary non-motile plants. They are, in fact, common to all plants.

A wider generalisation on the effects of Direct and Indirect Stimulation can now be established:

(1) THE EFFECT OF ALL FORMS OF DIRECT STIMULATION IS DIMINUTION OF TURGOR, CONTRACTION, DIMINISHED RATE OF GROWTH, A NEGATIVE MECHANICAL RESPONSE, AND AN ELECTRIC RESPONSE OF GALVANOMETRIC NEGATIVITY.

(2) THE EFFECT OF INDIRECT STIMULATION OF

FEEBLE OR MODERATE INTENSITY IS INCREASE OF TURGOR, EXPANSION, ACCELERATION OF THE RATE OF GROWTH, A POSITIVE MECHANICAL AND A POSITIVE ELECTRICAL RESPONSE.

SUMMARY

Under direct stimulation all plants and their different organs exhibit an excitatory reaction of galvanometric negativity.

The electric response disappears on the death of the tissue.

An impulse transmitted in a non-conducting or semi-conducting tissue induces an electric change of galvanometric positivity.

The electric response of the midrib to transmitted impulse is negative. Moderate stimulation enhances the conducting power, with resulting staircase-increase in successive responses. The conduction is irreciprocal. Under minimal stimulus conduction takes place only in a centrifugal direction: it is only under strong stimulus that conduction takes place in both directions.

When two electric connections are made on the midrib, the transmitted impulse gives rise to a diphasic response; this is due to the arrival of the excitatory wave first at the proximal contact, and later at the distal contact.

The maximum excitation at one of the contacts is attained in the course of about 0·8 second; it disappears in the further course of 4·3 seconds.

The electric responses of ordinary plants to direct and indirect stimulation are parallel to the mechanical responses of motile sensitive plants.

CHAPTER X

ELECTRIC RESPONSE OF MIMOSA TO INDIRECT STIMULATION

THOUGH the characteristic responses to indirect stimulation were shown in the last chapter to be universal in all plants, it is desirable to repeat the experiments with the sensitive Mimosa: for with this plant it is possible to obtain in one

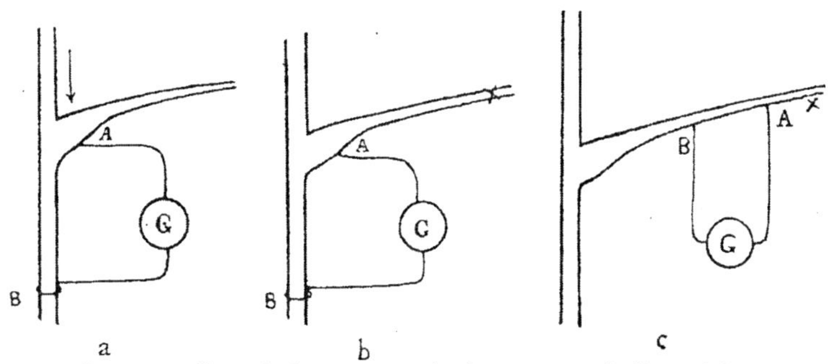

FIG. 43.—Electric Connections for Response to Indirect Stimulation in Mimosa.

a, Electric connection for response of lower half of pulvinus to stimulus applied on the upper half:
b, for Monophasic response of pulvinus to stimulus applied on the petiole at cross.
c, for Diphasic response of petiole to stimulus applied to petiole at cross.

experiment a record of both the mechanical and the electric responses, so as to demonstrate the practical identity of the two reactions. Thus, on application of sub-maximal stimulus to the sub-petiole, it was found that the mechanical and electrical responses occurred almost simultaneously. The very slight difference between the two is accounted for by the delay in starting the motile mechanism of the pulvinus into action.

The electrical connections for obtaining the electric response are given diagrammatically in fig. 43. A thin platinum wire is thrust about 1 mm. into the pulvinus, the second contact being made at a distant indifferent point on the surface of the stem, which is not reached by the excitatory impulse. The effect of stimulation transmitted across the semi-conducting pulvinus can be obtained with similar electric connections, stimulus being applied locally on the upper surface of the pulvinus. To record the velocity of the nervous impulse in the conducting strand of the petiole, two fine platinum wires are thrust in to make the necessary electric connections with the recording galvanometer.

ELECTRIC RESPONSE TO INDIRECT STIMULATION

It was ascertained in Experiment 48 that the mechanical response to feeble indirect stimulation was positive, which was transformed into negative under stimulus of a stronger intensity.

FIG. 44.

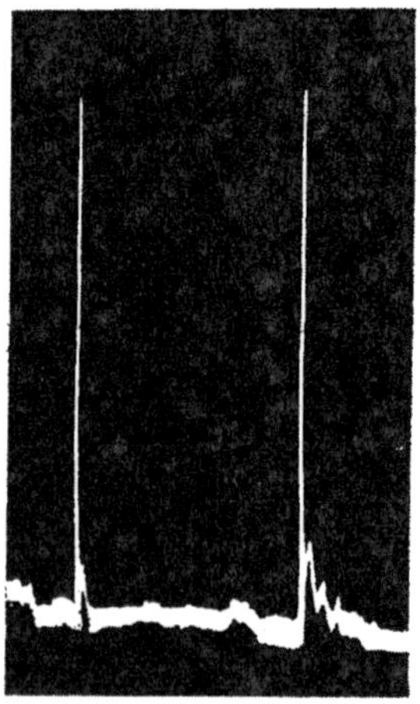

FIG. 45.

FIG. 44.—Positive Response (Mimosa) to feeble Stimulus.

FIG. 45.—Response of Galvanometric Negativity to stronger Stimulus.

Experiment 59. *Electric response to feeble stimulus.* — Records of response to feeble stimulus applied on the distant sub-petiole are given in fig. 44; the responses are seen to be positive, that is to say, the electric change is one of galvanometric positivity.

Experiment 60. *Electric response to strong stimulus.* — The experiment was repeated with the same specimen, the only difference being that the intensity of the stimulus was made sub-maximal. The response is seen to be an electric change of galvanometric negativity (fig. 45).

It may be thought that the electric variation might be due to the mechanical movement of the falling leaf which would occur under excitation. That this is not the case is proved by holding the leaf in a fixed position, when the electric response is found to be the same as before. The mechanical and electric responses are therefore independent manifestations of a common excitatory reaction.

The Excitatory and Contractile Processes

A muscle immersed in distilled water loses its power of contraction. Application of water to the pulvinus of Mimosa also renders it mechanically irresponsive. I have shown that the conductivity of the embedded nerve is not modified by the abolition of motile excitability. The question now arises whether the excitatory process is still in operation in the water-logged pulvinus, though it does not exhibit it by actual contraction. The problem was solved by the following experiment.

Experiment 61.—The pulvinus of Mimosa was made mechanically irresponsive by application of water. In this condition, stimulation, as already stated, is not followed by the normal fall of the leaf. An electric record of response was then obtained, which showed that the mechanically insensitive leaf gave the normal electric response of galvanometric negativity, proving that the tissue was still irritable though unable to manifest it outwardly by mechanical movement. This fact helps to explain the occurrence of excitatory electric response in ordinary plants in which mechanical response is practically absent.

Determination of Velocity of Excitatory Impulse by Diphasic Response

I next attempted the difficult task of recording the extremely short interval of the passage of the excitatory impulse from A to B in the conducting tissue itself. Two fine platinum wires made the necessary electric connections with the conducting phloem (fig. 43, c), stimulus being applied at a point not very distant from A. In this method, A is first rendered galvanometrically negative; the excitation then reaches B, rendering the point negative, and causing a reversal of the electric response. The interval between the two reactions is extremely short, specially when the two points A and B are separated by a comparatively short distance. The record of the diphasic response cannot therefore be obtained by an ordinary galvanometer; it can only be secured, as previously explained, by the string-galvanometer of Einthoven. Before describing the results obtained with the plant-nerve, I reproduce the record of diphasic response of the olfactory nerve of the Pike obtained by Garten with the Capillary Electrometer (fig. 46). The proximal point on the nerve first became negative; the later arrival of the impulse at the second contact is seen as a reversed response due to the negativity induced at that point.

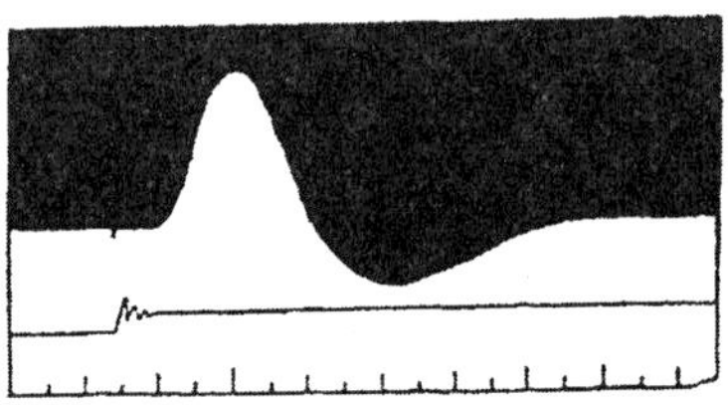

FIG. 46.—Diphasic Response in Olfactory Nerve of Pike as shown by the Capillary Electrometer (after Garten).
Signal for stimulus and time-marks given below.

Experiment 62.—I now describe the diphasic response of the nerve of Mimosa obtained with the Einthoven string-galvanometer. The two points of contact were at a distance of 7 mm. from each other. The arrival of the excitatory wave at A is signalled by an up-curve denoting galvanometric negativity at that point. The arrival of the wave at B produces a reversal of the curve due to the electric negativity at that point (fig. 47). The time-interval between the responses A and B is ·25 second; as the distance traversed by the impulse is 7 mm., the velocity of transmission of excitation is 28 mm. per second, which is practically the same as the value obtained by the mechanical method. In summer this varies in different specimens from about 15 to 30 mm. per second.

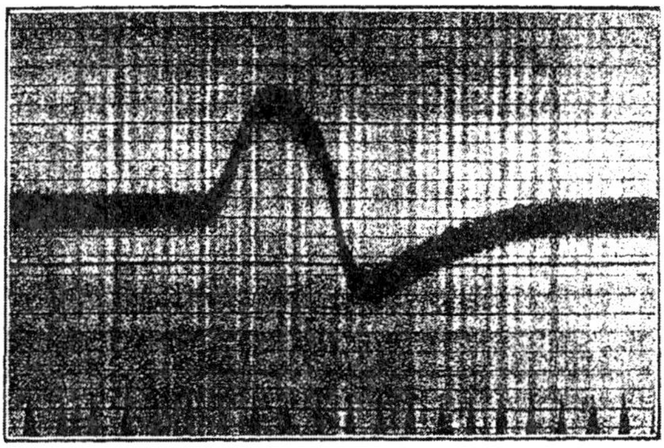

FIG. 47.—Diphasic Electric Response of Nerve of Mimosa; with Einthoven Galvanometer (see text).

I reproduce another record of the diphasic response of the conducting nerve of Mimosa; the sensitiveness of the

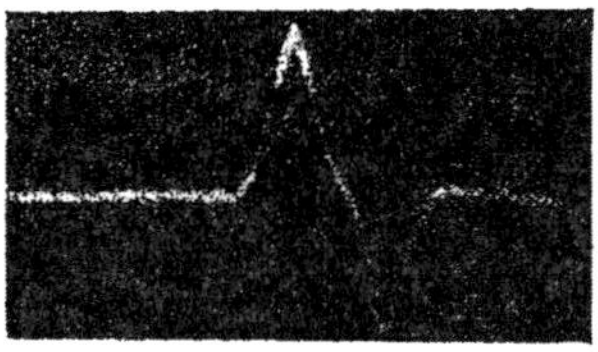

FIG. 47A. A second record of the diphasic response of nerve of Mimosa: obtained after fitting a finer string to the galvanometer.

string-galvanometer was increased, with resulting enhancement of the amplitude of response (fig. 47A). The distance between the two points of contact was 10 mm. and the time interval between responses of A and B was 0·5 second (time-marks not shown in the figure). This gives 20 mm. per second as the velocity of transmission of excitation.

Electric Localisation of the Nervous Tissue in Mimosa

It is impossible to isolate the nervous tissue in Mimosa without injury ; hence we are confronted with the problem of localising *in situ* the particular tissue which serves as the conductor of excitation. The problem was ultimately solved by the invention of the Electric Probe.

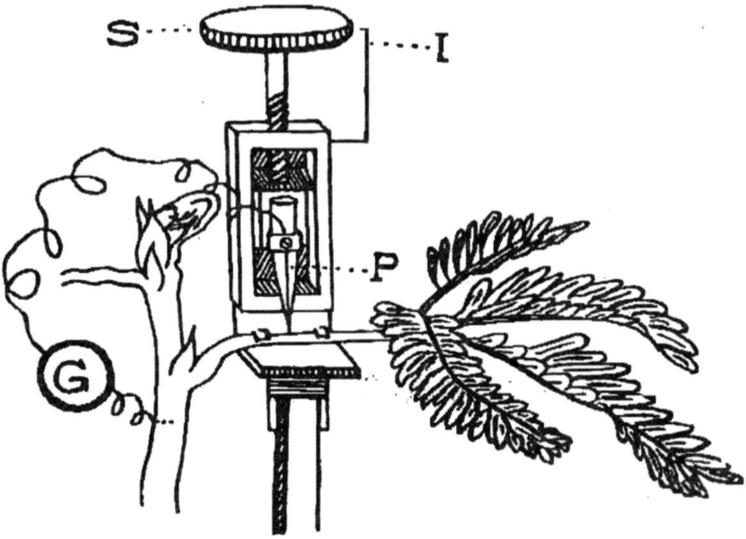

Fig. 48.—The Electric Probe for localisation of Nervous Tissue in the petiole of Mimosa.

P, the probe in circuit with the galvanometer, G ; S, the screw-head, by the rotation of which the probe is forced into the petiole ; I, index by which the depth of intrusion may be determined.

The principle of the method will be readily understood if the petiole be thought of as a cable along which electric messages are being transmitted. The conducting strand is here inside a non-conducting sheath. The embedded conductor can be localised, and the transmitted messages picked up, by gradually thrusting in the Electric Probe, which is insulated except at the extreme tip. A galvanometer included in the circuit of the probe indicates the messages that are being transmitted as soon as the tip of the Probe comes in contact with the conducting strand. The depth of insertion attained can be read on a suitable scale, and thus the position of the conductor can be determined.

The exact position of the conducting nerve embedded in the petiole of Mimosa can be localised in a similar manner. Excitation of the sub-petiole gives rise to an excitatory impulse which travels in a centrifugal direction, the impulse in the nerve being detected by a change to galvanometric negativity. Of all the tissues the conducting nerve will be most intensely excited by the transmitted impulse, and the induced electrical change of this particular tissue will be maximum. Excitation will no doubt be irradiated, but this will show a rapid diminution in the more external tissues. If the stimulus be moderate or feeble the irradiation will be slight.

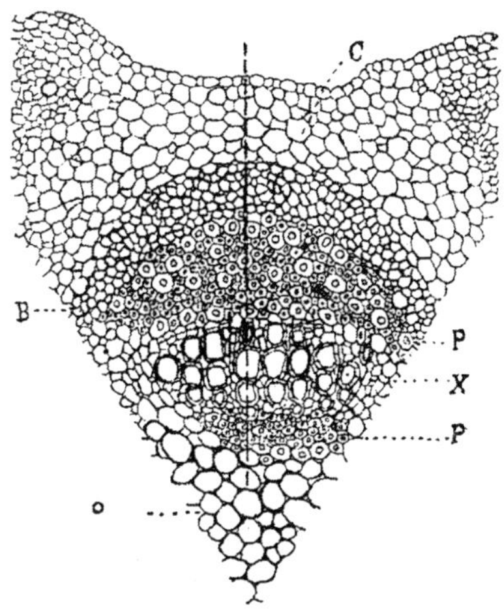

FIG. 49.—Transverse Section of the Upper Vascular Bundle containing Conducting Nerve.

Dotted line indicates line of passage of the Probe.

The experimental procedure is as follows: The probe is thrust in perpendicularly to the diameter of the petiole from above (fig. 48). The intrusion of the probe is by steps, say, of 0·05 mm. at a time. The slight wound produced by the insertion of the tip of the probe causes excitation, which subsides completely in the course of about fifteen minutes. The third sub-petiole in connection with the upper nerve is periodically stimulated by uniform stimuli.

Experiment 63. *Electrical excitation in the various*

tissues.—I now give detailed results of the localisation of the conducting tissue. The probe enters the epidermis and is pushed in by steps of 0·05 mm.; it passes in succession the cortex C, the outer phloem P, the xylem X, the inner phloem P′, and the central pith O. The thickness of the different tissues is modified by the age of the specimen.

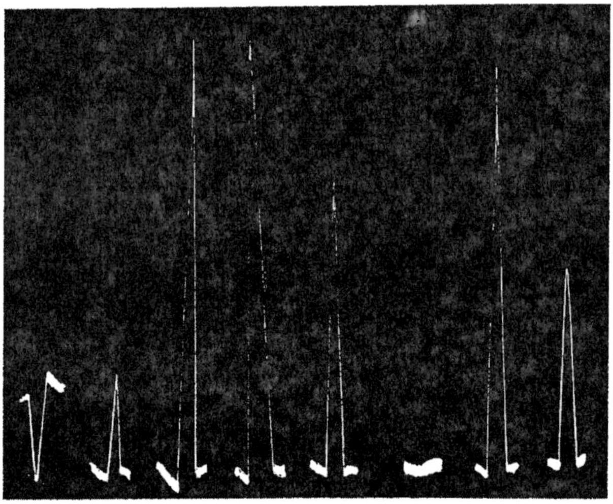

FIG. 50.—The Galvanometric Record of Transmitted Excitation in different tissues of the Petiole.

The first is the positive response of the epidermis; the second is the feeble negative response of the cortex; the third, fourth and the fifth are the enhanced responses in the outer phloem; the sixth shows absence of excitation in the xylem; the seventh is the enhanced response in the inner phloem; the eighth is the diminished response in the pith.

In the records given above (fig. 50) the electric response of the epidermis was + 12 divisions of the galvanometer. I have shown elsewhere that the epidermis, which is a non-conductor, gives either a zero or a positive response, in contradistinction to the normal negative response of conducting tissue. The probe at a depth of 0·1 mm. encountered the cortex, when the response on account of irradiation was but slight, being — 17 divisions. It next arrived at the region of the phloem which extends through 0·15 mm., the average depth being 0·2 mm. The response in this region underwent a sudden enhancement, as seen in the three responses — 61, — 65 and — 40 divisions. The xylem, which was at a depth of 0·3 mm., showed little or no response, proving that it was practically a non-conductor.

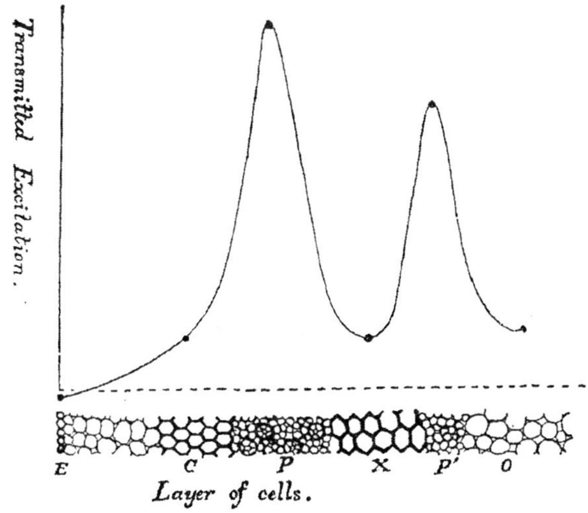

FIG. 51.—Curve showing the different intensities of Transmitted Excitation in the different tissues.

E, epidermis; C, cortex; P, outer phloem; X, xylem; P′, inner phloem; O, pith.

When the probe reached a depth of 0·35 mm. it encountered the inner phloem, and the response underwent a second enhancement of — 56 divisions. The probe reached the pith at a depth of 0·4 mm., and then the response underwent a diminution to — 26 divisions. In cases where the stimulation of the sub-petiole is feeble, the irradiation effects are greatly diminished; the excitatory transmission is then found only in the two phloems.

It will be seen that in all cases the phloem is invariably found to be the best channel for conduction of excitation, and that there are two conducting phloems, one external and the other internal. The curve (fig. 51), plotted from the mean values of ten different experiments, illustrates this in a striking manner. In the diagram of the transverse section usually given in text-books, there is in each bundle a single phloem-strand outside the xylem. The existence of two electrical maxima drew my attention to the possibility of there being two phloems in the bundle. As stated in a previous chapter, differential staining clearly brings out the fact that the phloem is not single but double (p. 35).

Impulse Transmitted across the Semi-Conducting Pulvinus

It was shown (p. 91) that the pulvinar tissue of Mimosa is a semi-conductor. It was also shown that local application of the stimulus of light on the upper surface of the pulvinus induced an increase of turgor and expansion, manifested by a preliminary positive or erectile response. The excitatory impulse subsequently reached the more excitable lower half and produced an abrupt negative response, the fall of the leaf (*cf.* fig. 35). Such is the mechanical response.

FIG. 52.—The Electric Response of the lower half of the pulvinus to impulse transmitted from the upper half.

Note preliminary positive (up-curve) followed by the more intense excitatory negative. Time-marks, one-tenth of a second.

Experiment 64.—I will now describe the corresponding electrical response. The electric connections will be understood from fig. 43 (a) in which the contact A is made with the lower half of the pulvinus, B being made with a distant indifferent point on the surface of the stem. Stimulation of the upper surface without mechanical disturbance was effected by application of a small drop of dilute hydrochloric acid. The electric response given by the Einthoven string-galvanometer is reproduced in fig. 52. It will be seen that there is an up-response indicating the brief galvanometric positivity at A due to the preliminary positive impulse in a semi-conductor. The excitatory impulse then reached A, and caused the more intense down-response of galvanometric negativity at that point. When the stimulation of the upper surface is intensified by the application of a more concentrated solution of hydrochloric acid, the velocity of the excitatory impulse is greatly increased, with the result that the more intense negative response masks the feeble preliminary positive. This proves once more that the transmitted impulse is double in every case, and that in highly conducting tissues the preliminary positive impulse becomes obliterated by the predominant negative which overpowers it.

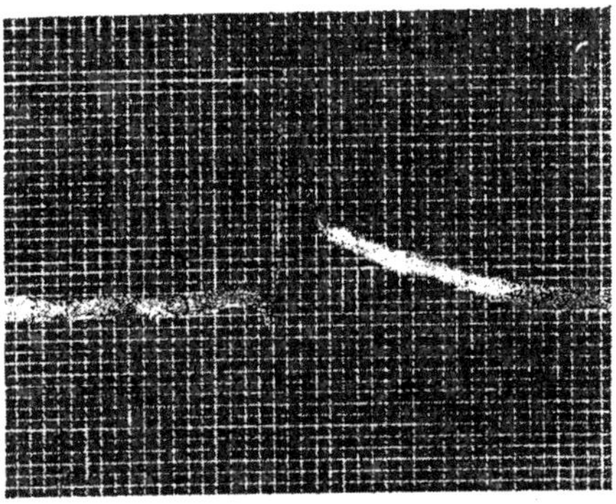

FIG. 52A. Electric record of positive and negative responses to indirect stimulation of the petiole of Mimosa: positive, down-curve.

Space between vertical lines = ·1 second.

One of the difficulties in obtaining a preliminary positive mechanical response of Mimosa to indirect stimulation of the conducting petiole is the inertia of the heavy leaf; two rapidly succeeding impulses of opposite sign cause a resultant movement which is the predominant negative. The two impulses positive and negative can, however, be recorded electrically by the Einthoven galvanometer, the inertia of which is negligible. Excitation by scratch-stimulus was effected on the lower side of the petiole, at a distance of 2 cm. from the pulvinus. The proximal electric connection was made with the pulvinus, and the distal with the indifferent surface of the stem. The record (fig. 52A) shows that the interval between the feeble positive and the stronger negative was only 0·05 second. A mechanical record of the preliminary positive impulse would, under the circumstances, have been an impossibility. The result given by the electric record is of importance, proving that two impulses (A and D reactions) are generated even in a conducting tissue, though the mechanical response fails to show the positive.

PRELIMINARY POSITIVE IMPULSE IN CONDUCTING TISSUES OF THE ANIMAL

The investigations on response in plants which I have been carrying out for the last twenty-five years have proved that the reactions in the plant are essentially similar to those in the animal. An apparent discrepancy arises from the fact that no indication of a preliminary positive response to the impulse transmitted along animal tissues had hitherto been noted.

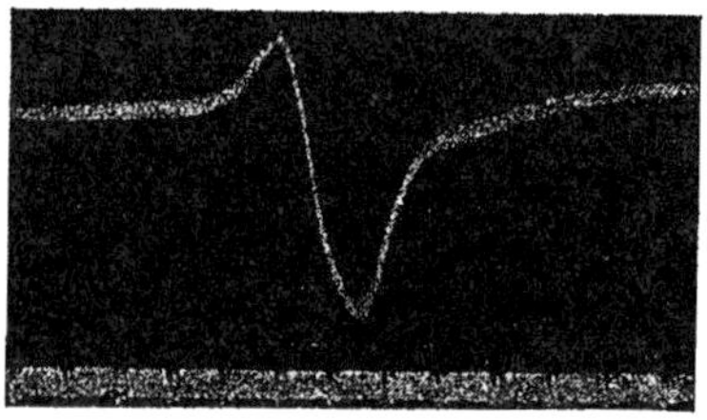

FIG. 53.—Electric changes accompanying the passage of a wave of contraction along the Ureter (Orbeli and Brucke): up-curve positive.

[The figure is taken from Bayliss—*General Physiology*, p. 653.]

I will now show that a preliminary positive response is given by animal tissues, which is often masked by the predominant negative. The positive can be unmasked by the employment of a semi-conducting tissue, and obtaining the record by an Einthoven galvanometer. The result is shown in the following Einthoven-galvanometer record (fig. 53) obtained by Orbeli and Brucke of the transmitted impulses in the semi-conducting ureter of the animal. The proximal electrode at which the wave first arrived became positive (up-curve); it then became strongly negative, due to the later arrival of the excitatory wave. The method is monophasic, the wave disappearing before arrival at the second electrode. As far as I am aware, no satisfactory explanation has been offered of this preliminary positive response. It finds a simple and satisfactory explanation if it be regarded as due to the positive impulse which precedes the negative.

Another method of obtaining the positive response, even in a highly conducting animal nerve, is to allow the tissue to fall into a condition of sub-tonicity, the nerve being thus rendered a semi-conductor or non-conductor. It will be shown in the next chapter that indirect stimulation gives rise to a positive impulse in both plant and animal nerve when sub-tonic, and that the restoration of full conductivity, as the result of repeated stimulation, is attended by transformation of the abnormal positive to the normal negative response.

It has been shown that the conducting power is not solely dependent on the nature of the tissue, but also on the energy absorbed by it from previous stimulation which raises its tonic level. The same tissue may, under different circumstances, be a conductor or a non-conductor of excitation. How, then, does stimulation modify the properties of a tissue? Is the modification produced only in conducting tissue, or in all tissues? If the latter, how are these modifications manifested?

Internal and External Work of Stimulus

Under favourable conditions the leaf of Mimosa responds to stimulation by a fall from a higher to a lower level of energy. Hence it is generally supposed that response to stimulus is always attended by a depletion of energy. It has, however, been shown that the response, instead of being of one kind, is of two kinds, a positive and a negative (*cf.* Chapter VIII.), whether the response be electric or mechanical. The positive erectile response with electrical positivity, in contrast to the negative response of fall with electrical negativity, indicates an accession of energy. The positive and negative responses, mechanical and electrical, are but external indications of fundamental reactions associated with storage or expenditure of energy. Work is done *on* the plant by the energy received from the environment, and work is done *by* the plant in the maintenance of its various life-activities. Representing the accession of energy by A, and the expenditure or depletion by D, the resultant effect is A − D, the difference between the up-hill and the down-hill work. The relative intensity of the two reactions to stimulation is, moreover, determined by the tonic condition of the tissue; when the tonic level is *above par*, the D-reaction under stimulus is predominant; but when the tonic level is *below par*, a condition designated as sub-tonic, the accession of energy A is the more pronounced, the tonic level being thereby raised to the normal. Turning to the protoplasmic property of conduction in a sub-tonic tissue, the resistance or block offered by the synaptic membranes is gradually removed by repeated stimulation, which is not unlike the removal of frictional resistance by the repeated working of a rusty hinge. The semi-conducting or non-conducting tissue thus becomes conducting after strong or repeated stimulation (p. 51).

PROTOPLASMIC IRRITABILITY AND ITS DIVERSE MANIFESTATIONS

The mechanical and electrical responses already described are not the only indications of protoplasmic reaction. The three fundamental manifestations of irritability—contractility, conductivity, and rhythmicity—are common to all living tissues. Evolution, associated with physiological division of labour, gives rise to tissue-systems specialised for the better discharge of particular functions. Nevertheless a continuity exists in all the different tissue-systems in respect of these three manifestations. Thus while the conductivity of the parenchymatous tissue is feeble, in the nervous tissue the tubular cells conduct excitation with little decrement to a considerable distance. Again, different grades of contractility are exhibited by different tissues; it is, for example, low in ordinary pulvini, but reaches high perfection in the pulvinar tissue of Mimosa. In regard to rhythmicity, the highest development is attained in the autonomous activities of the pulsating leaflets of *Desmodium gyrans* and of growing organs. The response of protoplasm to stimulation is given mechanically, by arrest or greater activity of movement or of conduction; electrically, by variation of electromotivity.

Parallel Effects of Stimulation on Contractility, Conductivity and Rhythmicity

The conclusion that a fundamental protoplasmic change is induced by stimulation is based upon observations which demonstrate that stimulation produces diametrically opposite reactions in tissues which are respectively in sub-tonic or in normal condition. The diverse manifestations of protoplasmic irritability are, as already explained, contractility, electromotivity, conductivity, and rhythmicity. In the condition of sub-tonicity, all these manifestations become depressed or arrested. The pulvinus ceases to contract, the electromotivity of the tissues becomes depressed, the conducting phloem ceases to conduct the excitatory impulse, and the pulsating activity of the leaflet of Desmodium and of the growing organ come to a stop. In optimum condition, when the tonicity is above par, the contractile and electromotive responses are at their maximum, the velocity of conduction attains its highest value, and the autonomic activities of Desmodium leaflet and of growing organs are at their greatest.

Effect of Stimulation on Sub-tonic Tissues.

Induced enhancement of motile excitability.—I have shown elsewhere[1] that the lost excitability of the pulvinus

[1] *Life Movements in Plants* (1918), p. 147.

of Mimosa becomes restored under repeated stimulation. The mechanical and electrical responses then exhibit a staircase increase.

Enhancement of conductivity.—The after-effect of stimulation on the enhancement of conductivity of sub-tonic tissue has been demonstrated by the methods of both mechanical and electric response (pp. 51, 101).

Restoration of autonomous activity.—The arrested pulsation of Desmodium leaflet becomes revived under the action of stimulus. Stimulation also revives arrested growth.

Effect of Stimulation on Tissues in Optimum Condition

The responsive reactions are the converse of those of the tissues when in sub-tonic condition. Motile excitability, conductivity, and rhythmic activity are all depressed by strong stimulation.

Stimulation thus induces parallel modifications in conductivity, contractility, and rhythmicity.

The Leaf as a Catchment-basin for Stimulus

There is a particular aspect of the action of stimulus which is of fundamental importance in the life of the plant. The continuance of its normal functions depends on the maintenance by external stimulation of all the tissues in an optimum tonic condition; for deprivation of stimulation reduces the plant to an atonic condition in which all life-activities are brought to a standstill. Beginning with the tissue at the lowest tonic level (due to prolonged deprivation of stimulation), the incidence of stimulus initiates and enhances all physiological activities to a maximum, the tonic condition of the tissue being raised at the same time to an optimum. Among the external stimuli naturally accessible to the plant, none is more potent than light. All the conditions favour the transmission of its stimulating effect to a distance by the nervous channel, which is the phloem in the vascular tissue. The expanded lamina of the leaf, in which the vascular bundles are spread out in fine ramifications, is not merely a specialised structure for photosynthesis, but also a catchment-basin for the stimulus of light, the excitatory effect of which is gathered into larger and larger nerve-trunks for transmission to the interior of the plant. The distribution of the vascular bundles in the interior is such that no mass of living tissue is too remote to be excited by the stimulus conducted by these nervous channels. How reticulated they may often be, even in the trunk of a tree, is shown in the photograph of the distribution of the vascular bundles in the main stem of Papaya (fig. 54). This net-work, of which only a small portion is seen in the photograph, girdles the stem throughout its whole length, and in this particular case there were as many as twenty such layers, one within the other.

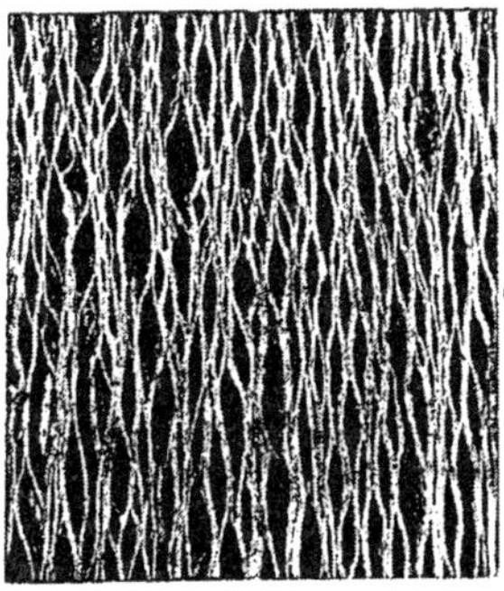

FIG. 54.—Distribution of vascular tissue in a single layer of stem of Papaya.

Thus all parts of the plant are maintained, by means of nerve-connection, in the most intimate and rapid communication with each other. It can only be in virtue of the existence of a system of nerves that the plant constitutes a single organised whole, each of whose parts is affected by every influence that falls upon any other.

SUMMARY

The electric responses of Mimosa to indirect stimulation and those of ordinary plants exhibit the closest parallelism.

The response to feeble stimulation is positive; under strong stimulation it is transformed into negative.

The excitatory process still persists in the pulvinus when rendered immotile by excessive absorption of water, as demonstrated by its excitatory response of galvanometric negativity.

The velocity of transmission of excitation can be determined electrically by the diphasic response obtained with the Einthoven galvanometer. The velocity thus determined is similar to that obtained by the mechanical method.

The nervous tissue in the petiole of Mimosa has been localised by the Electric Probe. The conducting tissue is the phloem: each vascular bundle contains two phloem-strands, one external and the other internal to the xylem.

In the semi-conducting pulvinus the impulses transmitted across it are double, positive followed by negative. Under strong stimulation the positive becomes masked by the predominant negative.

No preliminary positive response to transmitted impulse had hitherto been specifically noted in animal tissues. It has been shown that, under suitable conditions, positive response can be detected in these tissues.

The leaf is a catchment-basin for the stimulus of sunlight; the transmitted excitatory effect is conducted to the interior of the plant along the conducting phloem of the vascular bundles. It is the transmitted excitation that maintains the internal activities of the plant in optimum condition.

CHAPTER XI

RESPONSE OF ISOLATED PLANT-NERVE

THE characteristics of the electric response to transmitted excitation have been discussed in the last two chapters. It was shown that the phloem in the vascular bundle functions as the nerve of the plant. I had long been

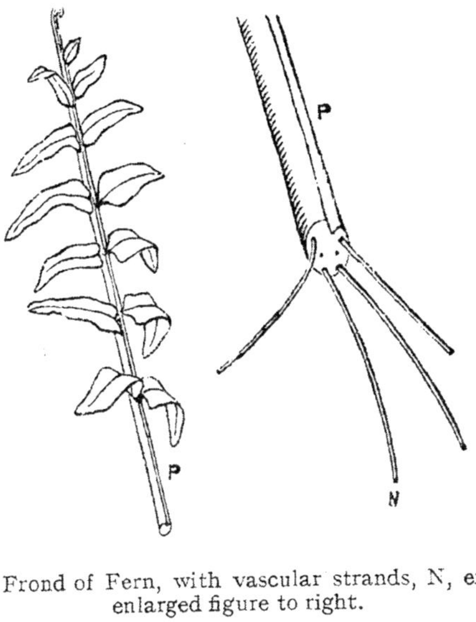

FIG. 55. Frond of Fern, with vascular strands, N, exposed in enlarged figure to right.

desirous of isolating the conducting or nervous tissue of the plant in order to repeat with it the characteristic experiments hitherto carried out with the nerve of the Frog. The petiole of the Fern afforded me the necessary material. On carefully breaking the hard casing of the petiole, and pulling it apart in both directions, I was able to isolate the vascular strands, which were long, soft, and white in colour, remarkably similar in their appearance to animal nerve (fig. 55). These threads vary in number with different species of Ferns; it is sometimes possible to detach one of them having a length of 20 cm. or more. In the vascular strand the conducting phloem surrounds the xylem. Transverse and longitudinal sections of a vascular strand are given in figure 56. The tubular conducting cells of the

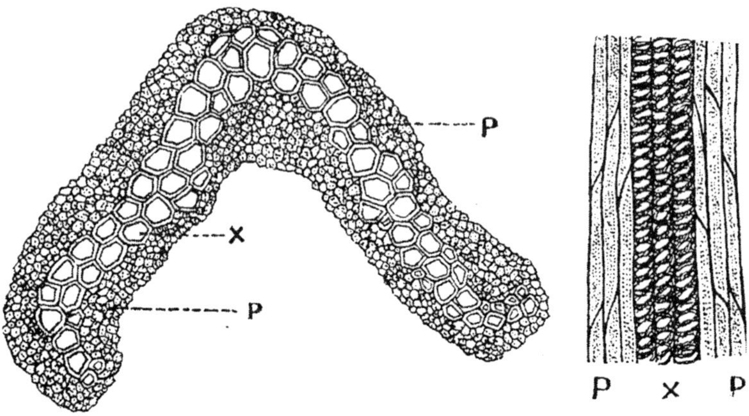

FIG. 56. Transverse and longitudinal sections of the vascular strand of a Fern.

P, conducting phloem; X, enclosed xylem.

phloem are similar to those in other plants. In the following investigations the strands of the common Maiden-Hair Fern (Adiantum) and of *Nephrodium molle* were found to be most suitable.

The vascular strand is dissected out, special care being taken to avoid injury. It is then placed in normal saline solution for about half an hour to remove all irritation due to handling. When the external temperature is not too high, the excitability of the isolated nerve-strand remains constant for a considerable length of time. The experimental precautions to be taken are precisely the same as those observed in corresponding experiments with animal nerve; that is to say, the specimen should be placed in a moist chamber, for in both cases the process of drying is found to induce a transient increase of excitability followed by a permanent abolition of responsiveness. In preparing an experiment, one end of the strand may be killed by the local application of hot salt-solution: the galvanometric connections are then made, one with the killed, and the other with the unkilled portion. In order to ensure that the electrical indication shall be a true responsive reaction, it is well to use a non-electrical form of stimulus One of the most perfect forms—as previously explained—is the thermal. With a good specimen a single thermal shock, lasting for less than a second, will be found sufficient to induce a considerable electrical response; a response of still greater amplitude may be obtained by the summated effects of several such stimulations.

An important characteristic of the animal nerve is its indefatigability. The responses to a long series of uniform stimuli, such as would in the case of ordinary tissues bring about marked fatigue, show little or none in that of the nerve. The same statement holds good of the plant-nerve: little fatigue is shown in a long-continued series of responses.

A complete account of the effect of chemical agents on the conductivity and excitability of the plant-nerve will be found in my 'Comparative Electro-Physiology,' 1907. I give here only a few of the more important results, specially in relation to the identity of the reactions given by the plant and by the animal nerve. This will be demonstrated by records which I obtained with the nerve of the Fern and with that of the Frog. Uniform thermal stimulation was employed for both.

Effect of Previous Stimulation on the Transmission of Impulse

The experiments on the electric response of the petiole-laminar preparation of ordinary plants and on the mechanical response of Mimosa, already described, show that :—

(1) The normal conductivity of the plant-nerve is enhanced in consequence of previous stimulation : excitation transmitted subsequently to stimulation is above the previous intensity, and the response to the transmitted excitation is above the normal. This fact is demonstrated by the staircase-increase of the electric response (see fig. 41). It is obvious that the continuous stimulation of tetanisation confers on the nerve a higher conductivity than at the beginning, the result of which must be a great enhancement of the electrical response.

(2) It has also been shown that after long isolation from the stimuli of its natural environment the tonic condition of the plant falls below par. The conductivity of the nerve, in such a sub-tonic specimen, is greatly depressed. The electric response of the sub-tonic nerve is the abnormal positive instead of the normal negative. Long continued stimulation restores the normal conductivity, the electric response to transmitted impulse being thereby converted from the abnormal positive to the normal negative.

I will now show that the above characteristics of response in normal and sub-tonic specimens are exhibited in a very striking manner by both the isolated nerve of the Fern and that of the Frog.

Effect of Tetanisation on the Conductivity of Nerve of Fern

Experiment 63. *Effect on normal nerve.*—The after-effect of tetanisation in enhancing the conducting power is exhibited in the records of figure 57. The first series of up-responses are the normal negative. The greatly increased amplitude of response to transmitted excitation after tetanisation indicates the increased conductivity of the nerve.

Experiment 64. *Effect on nerve in a sub-tonic condition.*—I reproduce (fig. 58) three types of response to transmitted impulse given by plant-nerve when normal and when in a depressed condition. In the three series, the first is

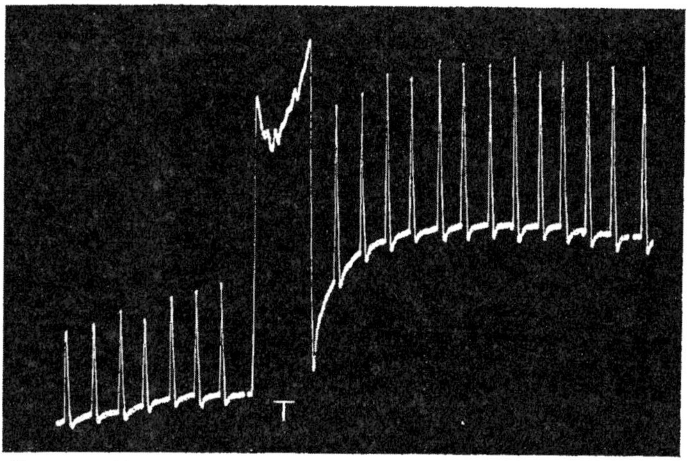

FIG. 57. Photographic record of effect of tetanisation, T, in inducing enhancement of normal negative response in nerve of Fern.

the normal negative; the second is the mixed or diphasic positive followed by negative; the third is the abnormal positive of a nerve in a condition of sub-tonicity.

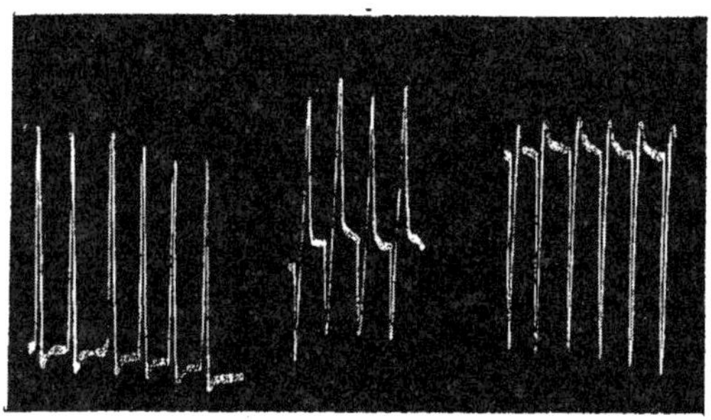

FIG. 58. Photographic record of three types of response, normal negative, diphasic, and abnormal positive, in nerve of Fern under different conditions.

In experimenting with the isolated nerve of the Fern, I find that the diphasic response is converted to the normal negative after a short period of tetanisation. The abnormal positive response requires a longer period of tetanisation for conversion into normal negative. This is accomplished, generally speaking, after an intermediate diphasic response in which the preliminary positive is followed by the negative.

These various effects find their natural explanation in the facts previously demonstrated that the block to the passage of impulse is gradually removed by previous stimulation. This is the reason why the conducting power of nerve in ordinary condition is raised above the normal, and why, when in the condition of sub-tonicity, the semi-conducting tissue is rendered normally conducting under similar treatment, the positive response being transformed into normal negative. I wish to lay special stress on these facts which will be found to offer the most natural and satisfactory explanation of corresponding phenomenon in the animal nerve.

Effect of Tetanisation on the Conductivity of Frog's Nerve

Experiment 65.—The responses were taken under conditions exactly parallel to those of the experiments with plant-nerve, the mode of stimulation being thermal. The first three responses are normal negative; the responses after tetanisation exhibit an enhancement similar to that observed in plant-nerve (fig. 59).

Experiment 66.—The Frog's nerve in sub-tonic condition exhibits, like the plant-nerve in a similar condition, a response of galvanometric positivity. After tetanisation the abnormal positive is found to be transformed into normal negative (fig. 60).

In his records of electrical response with Frog's nerve under electrical stimulation, Dr. Waller obtained responses of three different types. The first of these was the normal negative response; the second was the diphasic; and the third was the abnormal positive. This last he regarded as characteristic of stale nerve. The normal negative response was found by him to undergo enhancement after tetanisation, while the abnormal positive response of the stale nerve underwent a change into diphasic, or reversal to normal negative.

From the fact that carbonic acid enhances the normal negative response of nerve, Dr. Waller suggested that the

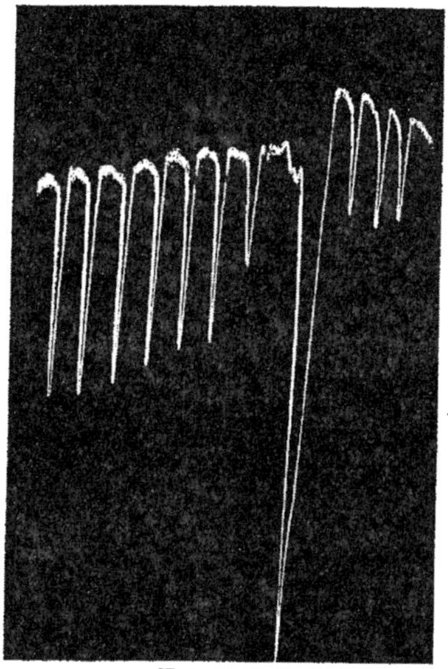

FIG. 59.

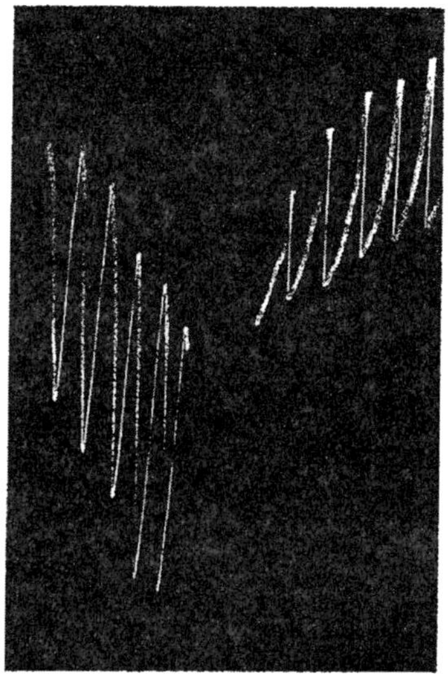

FIG. 60.

FIG. 59. Enhancement of amplitude of response, as after-effect of thermal tetanisation, in Frog's nerve.

The first three responses are normal. Brief thermal tetanisation is then applied, and the responses subsequently obtained under the original stimulation are enhanced.

FIG. 60. Conversion of abnormal positive (left) into normal negative response after thermal tetanisation (Frog's nerve).

enhancement of the normal negative after tetanisaiton and the tendency of the modified nerve to revert to the normal are results of a hypothetical evolution of carbonic acid in the nervous substance, due to metabolism accompanying the excitatory reactions.

It must be said, however, that the production of carbonic acid in the nerve is generally regarded as extremely doubtful.

Since it has been shown that all these varied effects are simply explicable on the well-demonstrated fact that conductivity becomes enhanced by the gradual removal of the block under stimulation, it is quite unnecessary to assume any hypothetical evolution of carbonic acid in the excited nerve.

The important generalisation which I have endeavoured to establish is the identity of the physiological mechanism in the plant- and in the animal-nerve. This finds complete demonstration in the similarity of reaction not only of normal nerves, but also of modified nerves.

Receptivity, Conductivity and Responsivity

When any point in the conducting tissue is acted on by an external stimulus, it receives the stimulation and is thrown into a state of excitation. This excitation is then conducted along the length of the tissue, and may be made outwardly manifest at some distant point by means of a suitable indication such as a mechanical or an electric response. Thus three distinct properties of the organism are called into play: namely, first, the sensitiveness at the point of reception of the stimulus, which I designate Receptivity; secondly, the power of transmission of the excitation, Conductivity; and thirdly, the manifestation of an excitatory effect at the distant responding region, which I term Responsivity.

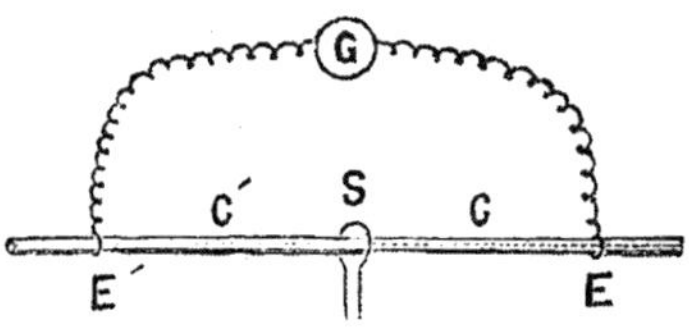

Fig. 61. Diagrammatic representation of the Conductivity Balance.

S, thermal stimulator; C and C′, the conducting arms of the balance; E and E′, responding points. Differential excitatory electrical effects at E and E′ recorded by galvanometer, G.

The measurement of variation of these under change of external conditions is attended by numerous difficulties, which have ultimately been overcome by my device of the Conductivity Balance (fig. 61). When the stimulus, given at S, is of sufficient intensity, the excitatory wave travels along both arms of the balance, through the conducting regions C and C′, and induces excitatory electromotive effects at the two responsive points E and E′. The excitatory electrical effects at E and E′ are opposed; and when they are equal they balance each other; and then the resulting galvanometer-deflection is reduced to zero.

Determination of Variation of Conductivity due to Chemical Agents

Detailed account of investigations on receptivity and responsivity will be found in my work already referred to. I will here describe some of the results of the determination of variation of conductivity induced by the action of chemical agents. The balanced record is taken at the beginning; the given chemical agent is then applied on a length of about 1 cm. at C on the right conducting arm (see fig. 61). If the effect of the agent is an increased conductivity of C, then the excitation transmitted to the right hand side E will be greater, and the upsetting of the balance will cause a resulting response upwards. Conversely a down-response will indicate that the effect of the agent has been to depress the conductivity. Again, it is possible to compare the relative effects on conductivity-variation induced by two different agents which are applied simultaneously, one on the arm C, and the other on C′.

Experiment 67.—As a typical example I describe the effect of dilute (·5 per cent.) solution of Na_2CO_3 on the conductivity of the nerve of the Fern. Inspection of figure 62 shows that a great and immediate enhancement of conductivity was produced, causing resultant up-responses which were particularly marked during the first four minutes. This increased conductivity then underwent continuous decrease and reversal into growing depression, as shown by the increasing down-responses. This record deserves special attention, inasmuch as it affords an insight into a phenomenon which could not otherwise have been suspected. Greater conductivity is usually associated with increased velocity of transmission. It would, however, appear that the term conductivity really covers two different phenomena which may not always be concomitant. That is to say, an increase of conductivity may mean either a greater speed of transmission of excitation, or a greater intensity of the excitation transmitted. The first four responses (fig. 62) show the induced enhancement of con-

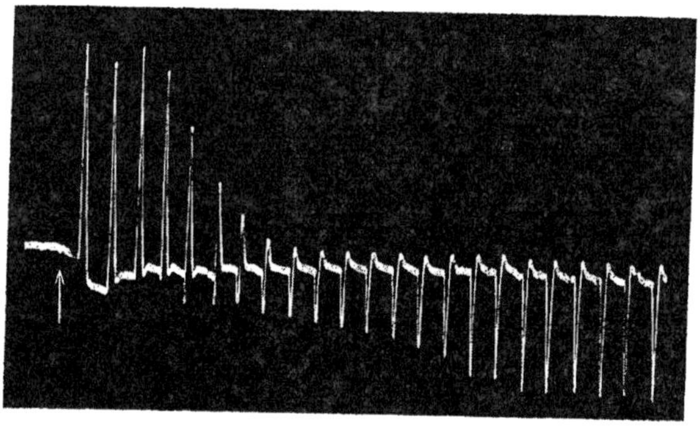

FIG. 62. Photographic record of effect of dilute (·5 per cent.) solution of Na_2CO_3 on conductivity (Fern-nerve).

ductivity by the fact that they are up-curves only. The fifth response, however, shows a marked preliminary twitch in the downward direction, followed by an up-response of some amplitude. This shows that the excitatory effect reached the right end E later than the left end E′, though the intensity still remained greater. The continued action of the reagent subsequently reduced the intensity also, so that this diphasic response ultimately became converted into a purely monophasic down-response.

The method thus provides an unique means of discriminating between the two distinct elements in conduction, namely, speed and intensity.

Effect of a Constant Current on Conduction

Experiment 68.—A special investigation was next undertaken of the effect of a constant electrical current in modifying normal conduction of excitation. The results obtained show that, under a feeble current, conduction is enhanced in an electric up-hill direction, that is to say, from the region of low to one of high electric potential; and conversely, that it is retarded or abolished in a down-hill direction, *i.e.* from a region of high to one of low potential. Further discussion of this totally unexpected result is given in the next chapter.

Summary

The vascular strands of the Fern can be readily isolated. They include the phloem which is the tissue conducting excitatory impulse, and which, like that of Mimosa, consists of long tubular cells. The isolated strand can be used experimentally, and may be described as a nerve.

A transmitted impulse gives rise, in the isolated nerve of the Fern, to a response of galvanometric negativity similar to that given by animal nerve.

Tetanisation enhances the conducting power in both plant and animal nerve.

When the nerve is in the condition of sub-tonicity, the response to a transmitted impulse is positive. This abnormal positive is transformed, after tetanisation, into the normal negative.

The characteristic responses of plant-nerve and animal-nerve are in every way similar, in both the normal and the modified conditions. Tetanisation has the same effect on both. These common characteristics prove that the physiological mechanism of excitatory conduction is the same in the plant and in the animal.

The Conductivity Balance enables accurate determination to be made of the effect of chemical agents on the conductivity of the nerve.

A constant electric current of feeble intensity enhances conduction in an electric up-hill direction; conversely, conduction is depressed or abolished in an electric down-hill direction.

CHAPTER XII

THE ELECTRIC CONTROL OF NERVOUS IMPULSE

A PROBLEM of great interest, which has attracted my attention for several years, is the question whether excitation travels better in a conducting tissue with or against the direction of an electric current passing through it. The definite results obtained with my Conductivity Balance on conduction in the isolated nerve of the Fern were so unexpected that I undertook to reinvestigate the subject by a different method, which has given satisfactory results. The investigation was carried out not only with the conducting tissue of Mimosa, but also with the nerve of the Frog.[1]

METHOD OF EXPERIMENT

The velocity of transmission in Mimosa is found from the mechanical response of the leaf registered by my Resonant Recorder, the writer being tuned to inscribe ten dots per second. After taking the record for normal transmission, two other records are obtained, with an electric current flowing along the petiole either with or against the direction of transmission of excitation.

The experimental arrangement is shown diagrammatically in fig. 63. After attaching the petiole to the recording lever, indirect stimulation is applied, generally speaking, at a distance of 15 mm. from the pulvinus. The stimulus of electric shock is given by means of a sliding induction-coil: the intensity of the shock is adjusted by

[1] 'The Influence of Homodromous and Heterodromous Current on Transmission of Excitation in Plant and Animal,' *Proc. Roy. Soc.*, B. Vol. 88, 1914.

gradually varying the distance between the secondary and the primary coils, till a minimally effective stimulus is

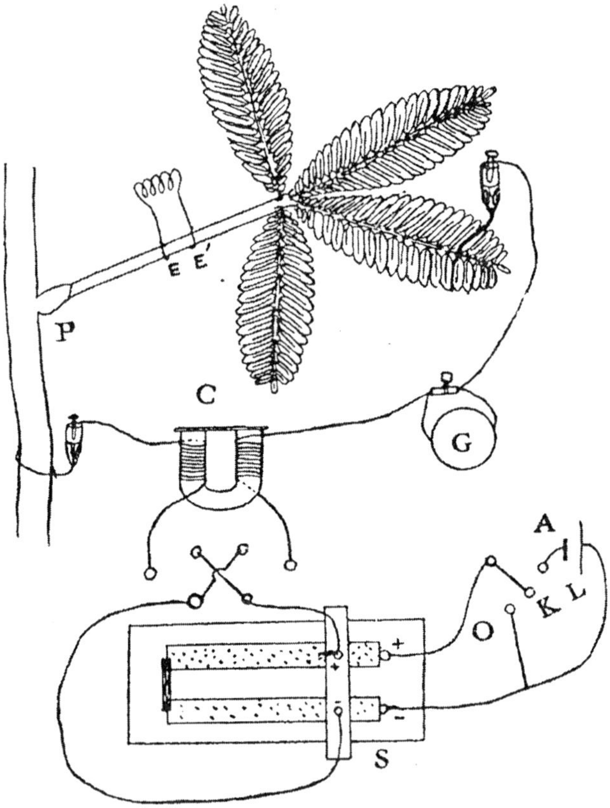

FIG. 63. Complete apparatus for investigation of the variation of conductivity in Mimosa.

A, storage-cell; S, potentiometer-slide, which, by alternate movement to right or left, continuously increases or decreases the E.M.F. of the constant current; K, switch-key for putting constant current 'on' and 'off' without variation of resistance; E, E', electrodes of induction-coil for stimulation; C, choking coil; G, micro-ammeter.

found. The constant current is conveyed to the plant by non-polarisable electrodes placed one on the stem and the other on the tip of a sub-petiole, at a distance from each other of about 95 mm. The point of stimulation and the responding pulvinus are both situated at a considerable distance from the anode or the kathode, in the indifferent region in which there is no polar variation of excitability. By means of a Pohl's commutator or reverser, the constant current can be maintained either 'with' or 'against' the direction of transmission of excitation. The transmission in the former case is 'down-hill,' and in the latter case 'up-hill.' Electrical connections are so arranged that when the commutator is tilted to the right, the transmission is down-hill, when tilted to the left, up-hill.

The electrical resistance offered by the 95 mm. length of stem and petiole was found to be from two to three million ohms. The intensity of the constant current flowing through the plant can be read by unplugging the key which short-circuits the micro-ammeter G. The choking coil C prevents the alternating induction-current from flowing into the polarising circuit, and causing direct stimulation of the pulvinus.

Before describing the experimental results, it is as well to enter briefly into the question of the external indication by which the conducting power may be gauged. A change of conductivity may be expected to give rise to a variation in the rate of propagation or to a variation in the magnitude of the excitatory impulse that is transmitted. There are thus several methods available for determining the induced variation of conductivity. In the first place, the variation of conductivity may be measured by the induced change in the velocity of transmission of excitation. In the second place, the transmitted impulse will give rise to an enhanced or a diminished amplitude of mechanical response determined by the increase or decrease of conductivity brought about by the directive action of the current. And, finally, the enhancement or depression of conductivity may be demonstrated by an ineffectively transmitted stimulus becoming effective, or an effectively transmitted stimulus becoming ineffective.

Effect of Direction of Current on Velocity of Transmission

A very convincing method of demonstrating the influence of a constant electric current on conductivity consists in the determination of the changes induced in the velocity of transmission by the directive action of the current. It is necessary to ascertain the true time required for the excitation to travel through a given length of the conducting tissue (1) in the absence of the current, (2) 'against,' and (3) 'with,' the direction of the current. The true time is obtained by subtracting the latent period of the pulvinus from the observed interval between stimulation and response. Now the latent period may not remain constant, but may undergo change under the action of the polarising current. It has been stated that the excitability of the pulvinus does not undergo any change when it is situated in the middle or indifferent region. The following results prove that, under parallel conditions, the latent period also remains unaffected.

Table IV.—The Effect of Constant Electric Current on the Latent Period (Mimosa).

Specimens	I	II
Latent period under normal conditions .	0·10 sec.	0·09 sec.
Latent period under current from right to left	0·11 sec.	0·10 sec.
Latent period under current from left to right	0·09 sec.	0·09 sec.

These results (Table IV.) show that a current applied under the given conditions has practically no effect on the latent period, the slight variation being of the order of one-hundredth part of a second. This is quite negligible when the total observed transmission-time is, as in the following cases, equal to nearly 2 seconds.

INDUCED CHANGES IN THE VELOCITY OF TRANSMISSION

Having found that the average value of the latent period in summer is 0·1 sec., the determination of the influence of the direction of the current on velocity can be undertaken.

Experiment 69.—In this and in the following experiments the stimulus of induction-shock was, as a rule,

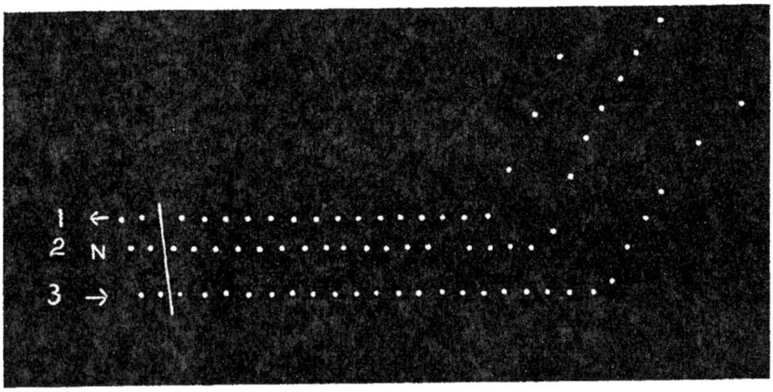

FIG. 64. Record showing enhancement of transmission 'up-hill' or against the constant current (uppermost curve) and retardation of velocity 'down-hill' or with the current (lowest curve).

N, normal record in the absence of current. ← indicates 'up-hill' and → 'down-hill' transmission.

applied on the petiole at a distance of 15 mm. from the pulvinus. The recording writer was tuned to 10 vibrations per second; the space between two succeeding dots, therefore, represents a time-interval of 0·1 second. The middle record, N in fig. 64, is the normal. There are seventeen spaces between the application of stimulus and the beginning of response. The total time is therefore 1·7 seconds, and by subtracting from it the latent period of 0·1 second, the true time, 1·6 seconds, is obtained. The normal velocity is found by dividing the distance 15 mm. by the true time 1·6 seconds: thus $V = 15 \div 1{\cdot}6 = 9{\cdot}4$ mm. per second.

With regard to the effect of current in modifying the normal velocity, the uppermost record (1), fig. 64, was taken under the action of an 'up-hill,' or 'against,' current of the intensity of 1·4 micro-amperes. It will be seen that the time was reduced from 1·7 seconds to 1·4 seconds; making allowance for the latent period, the velocity of transmission under 'up-hill' current $V_1 = 15 \div 1{\cdot}3 = 11{\cdot}5$ mm. per second. The lowest record (3) gives the effect of a 'down-hill' current; the time between stimulation and response was prolonged to 1·95 seconds and the velocity reduced to 8·1 mm. per second. The conclusion arrived at from the results of this mechanical mode of investigation is identical with that based upon the results obtained by the electric method of the Conductivity Balance referred to previously: it is as follows:—The passage of a feeble constant current modifies conductivity for excitation in a selective manner: it enhances conductivity when the direction of the constant current is opposite to that in which the excitation is travelling; it diminishes it when the direction of the constant current is the same as that in which the excitation is travelling. The results obtained with four different specimens are given in Table V.

The minimum current which induces a perceptible change of conductivity varies somewhat in different specimens. The average value of the minimal current in autumn is 1·4 microamperes. The effect of even a feeble current may be detected by employing a test-stimulus which is barely effective.

TABLE V.—SHOWING EFFECTS OF UP-HILL AND DOWN-HILL CURRENT OF FEEBLE INTENSITY ON TIME OF TRANSMISSION THROUGH 15 MM.

Number	Intensity of current in microamperes	Time for up-hill transmission	Time for down-hill transmission
1	1·4	14 tenths of a second	16 tenths of a second
2	1·4	13 ,, ,,	15 ,, ,,
3	1·7	12 ,, ,,	14 ,, ,,
4	1·6	19 ,, ,,	Arrest

Having demonstrated the effect of the direction of current on the velocity of transmission, I go on to describe other methods by which induced variation of conductivity may be detected.

The Method of Minimal Stimulus

In this method a minimal stimulus is employed, such that the transmitted impulse gives rise, under normal conditions, to a feeble response. If the passage of a constant current in a given direction enhances conductivity, then the intensity of the transmitted excitation will also be enhanced; the minimal response will tend to become maximal; excitation which had hitherto been ineffectively transmitted will now become effectively transmitted. Conversely, depression of conductivity will result in a diminution or abolition of response. A single break-shock of sufficient intensity may serve as the test-stimulus; or the additive effect of a definite number of induction-shocks, the alternating elements of which are exactly equal and opposite, which is secured by causing rapid reversal of the primary current by means of a rotating commutator.

Experiment 70.—Working in this way, it was found that an impulse transmitted against the direction of the constant current became effective or was enhanced. A current, flowing in the same direction as that of the transmitted impulse, on the other hand, diminished the intensity of transmitted excitation or blocked it altogether.

It will be convenient to designate the currents in the two directions; the current in the direction of transmission will be termed *Homodromous*, and that against the direction of transmission *Heterodromous.*

After-Effect of Homodromous and Heterodromous Currents

The passage of a current through a conducting tissue in a given direction causes, as has been shown, enhanced conductivity in an opposite direction. It may be supposed that this is brought about by a particular molecular arrangement induced by the current, which assists the propagation of the excitatory disturbance in a selected direction. On the cessation of this inducing force, there will be a rebound and a temporary reversal of the previous

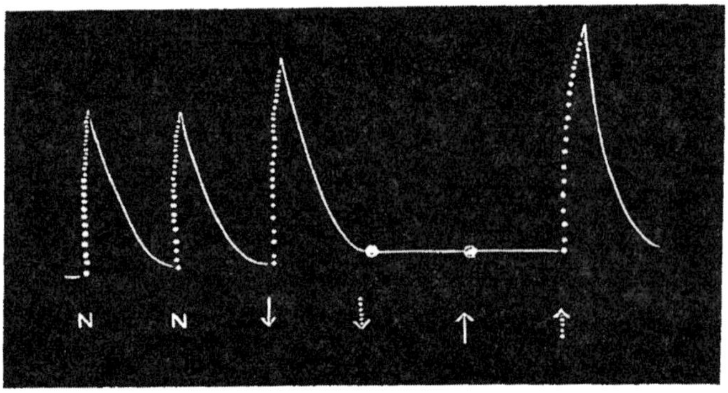

FIG. 65. Direct and after-effect of heterodromous and homodromous currents.

First two records, N, N, normal; ↓, enhanced transmission under heterodromous current; ↓, arrest of conduction as an after-effect of heterodromous current. Next record ↑ shows arrest under homodromous current. Last record ↑ shows enhancement of conduction greater than normal, as an after-effect of homodromous current. (Dotted arrow indicates the after-effect on cessation of a given current; ↑ homodromous and ↓ heterodromous current.) (Mimosa.)

molecular arrangement, with concomitant reversal of the conductivity-variation. The immediate after-effect of a current flowing in a particular direction on conductivity is likely to be a transient change, the sign of which would be opposite to that of the direct effect. The after-effect of a heterodromous current may thus be a temporary depression, that of a homodromous current a temporary enhancement of conductivity.

Experiment 71.—This anticipation is fully justified in the following experiment:—The first two responses were normal, after which the passage of a heterodromous current gave rise to an enhanced response. The depressing after-effect of the heterodromous current rendered the next stimulation ineffective. The next record, taken during the passage of the homodromous current, exhibits an abolition of response due to an induced depression of conductivity; and the last record, the after-effect of the homodromous current as a response larger than the normal (fig. 65). These results show that the after-effect of cessation of a current in a given direction is a transient conductivity-variation of which the sign is opposite to that induced by the continuation of the current.

Influence of Direction of Constant Current on the Conductivity of Animal Nerve

I now take up the question whether an electric current induces any selective variation of conductivity in the animal nerve, similar to that observed in the conducting tissue of the plant.

In the experiments which I am about to describe, special arrangements were made such that (1) the excitation had not to traverse the polar region, and that (2) the point of stimulation was at a relatively great distance from either pole. The fulfilment of the latter condition ensured the point of stimulation being placed in the neutral region.

In the choice of experimental specimens I was fortunate enough to secure Frogs of an unusually large size, locally known as 'Golden Frogs' (*Rana tigrina*). A preparation was made of the spine, the attached nerve, the muscle and the tendon. The electrodes for the constant current were applied at the extreme ends, on the spine and on the tendon (fig. 66). The following are the measurements, in a typical case, of the different parts of the preparation. Length of spine between the electrode and the nerve = 40 mm.; length of nerve = 90 mm.; length of muscle = 50 mm.;

length of tendon = 30 mm. Stimulus was applied in all cases on the nerve, at a point midway between the two

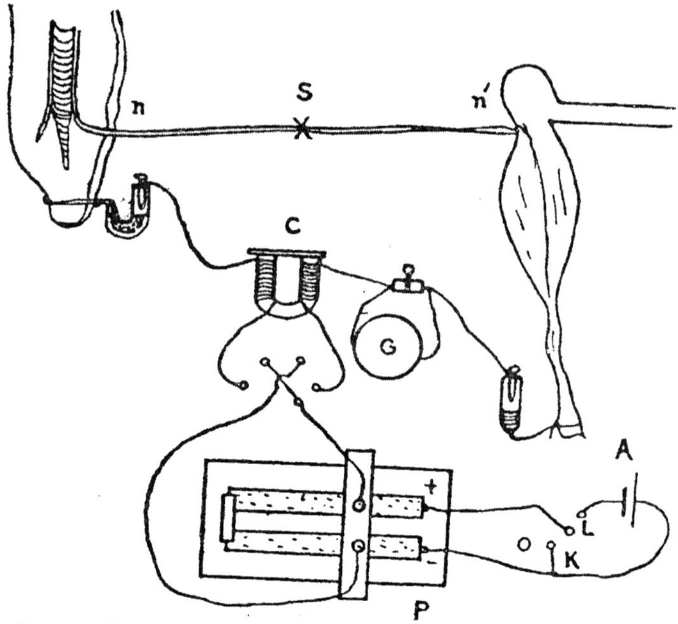

FIG. 66. Experimental arrangement for study of the effect on conductivity of Frog's nerve of the directive action of an electric current.

N, N′, nerve ; S, point of application of stimulus in the middle or indifferent region.

electrodes, and at a minimum distance of 100 mm. from either electrode. The point of stimulation is, therefore, situated in the indifferent region.

INDUCED VARIATION OF VELOCITY OF TRANSMISSION IN ANIMAL NERVE

Striking evidence of the influence of the direction of a constant current on conductivity in plants was afforded by the induced variation of the velocity of transmission.

Equally striking is the result which I have obtained with the nerve of the Frog.

Experiment 72.—The experiments described below were carried out during the cold weather. The following records (fig. 67), obtained by means of the pendulum-

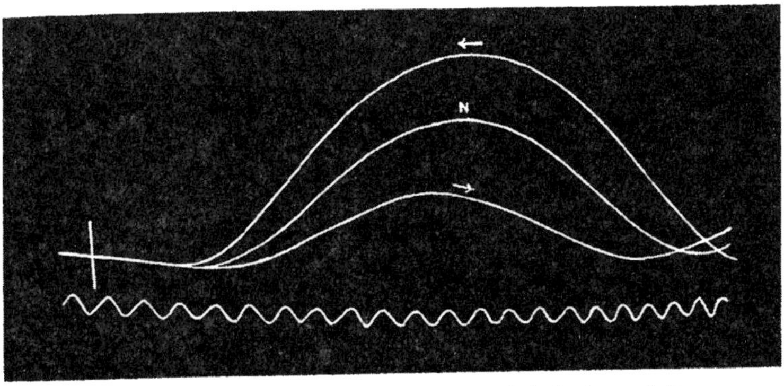

FIG. 67. Effect of constant electric current in inducing variation in velocity of transmission in animal nerve (Frog).

N, normal record; upper record shows enhancement, and lower record retardation of velocity of transmission under heterodromous and homodromous currents respectively.

myograph, exhibit the effect of the direction of the constant current on the time of transmission through a given length of nerve. The latent period of muscle being constant, the variations in the records exhibit the actual changed rates of conduction. The middle record is the normal, in the absence of any current. The upper record, denoted by the left-hand arrow, shows the action of a heterodromous current in shortening the transmission-time and thus enhancing the velocity above the normal rate. The lower record, denoted by the right-hand arrow, exhibits the effect of a homodromous current in retarding the velocity below the normal rate. I find that a very feeble heterodromous current is enough to induce a considerable increase of velocity, which soon reaches a limit. For inducing retardation of velocity, a relatively strong homodromous current is necessary. I give on p. 146 a table showing the results of several experiments.

TABLE VI.—EFFECT OF HETERODROMOUS AND HOMODROMOUS CURRENT OF FEEBLE INTENSITY ON VELOCITY OF TRANSMISSION.

Specimen	Intensity of heterodromous current	Acceleration above normal	Intensity of homodromous current	Retardation below normal
	microampere	per cent.	microamperes	per cent.
1	0·35	16	1·0	20
2	0·7	13	1·5	19
3	0·8	18	2·0	14
4	0·8	11	2·0	13
5	1·0	18	2·5	12
6	1·5	15	3·0	40

VARIATION OF INTENSITY OF TRANSMITTED EXCITATION

In the next method of investigation, the induced variation of intensity of transmitted excitation is inferred from

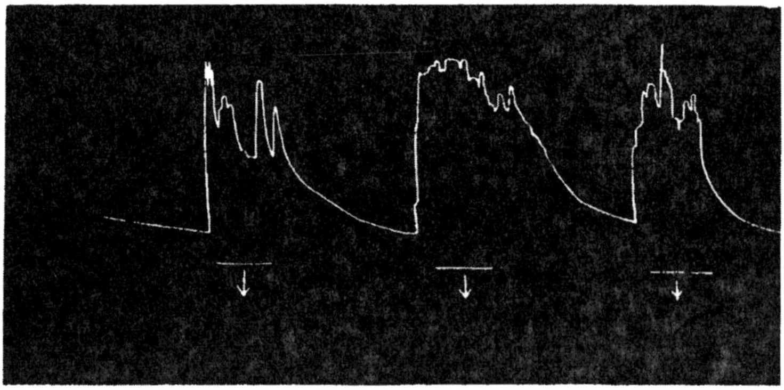

FIG. 68. Ineffectively transmitted salt-tetanus becoming effective under heterodromous current, denoted by down-pointing arrow (nerve of Frog).

the varying amplitude of response of the terminal muscle. Testing stimulus of sub-maximal intensity is applied at the middle of the nerve, where the constant current induces no variation of excitability. Stimulation is effected either by a single break-shock or by the summated effects of a definite number of equi-alternating shocks, or by chemical stimulus.

Experiment 73.—Under the action of feeble heterodromous current the transmitted excitation was always enhanced, whatever the form of stimulation: this is illustrated in fig. 68. Homodromous current, on the other hand, inhibited or blocked excitation (fig. 69).

After-effect of Heterodromous and Homodromous Currents

On the cessation of current there is induced in the plant-tissue a transient conductivity-change of opposite sign to that induced by the current (*cf.* Experiment 71). I find this to be the case also in animal nerve. I give a typical illustration of the direct- and after-effect of homodromous current on salt-tetanus.

Experiment 74.—In this experiment a sufficient length of time was allowed to elapse after the application of the salt to the nerve, so that the muscle, in response to the transmitted excitation, exhibited an incomplete tetanus T. The homodromous current was then applied, with the result of inducing a complete block of conduction and a concomitant disappearance of tetanus. The homodromous current was gradually reduced to zero by appropriate movement of the potentiometer-slide. The after-effect of the homodromous current was then shown in a transient enhancement of transmitted excitation, which lasted for nearly 40 seconds. After this the normal conductivity was restored. Repetition of the experiment gave similar results (fig. 69).

The results that have been given are typical of a very large number of experiments which invariably presented the characteristic phenomena that have been described. Thus it is established that, with feeble or moderate current, conductivity is enhanced when against the direction of the current, and depressed or blocked when with the direction of the current. Under stronger current the normal effect undergoes a reversal.

It has thus been shown that a perfect parallelism exists in the conductivity-variations induced in the plant and in the animal by the directive action of a constant current. No explanation can be regarded as satisfactory which is not applicable to both cases. Now with the plant it is possible to arrange the experimental conditions in such a way that the factor of variation of excitability is completely eliminated. The various effects manifested by the plant-tissue are

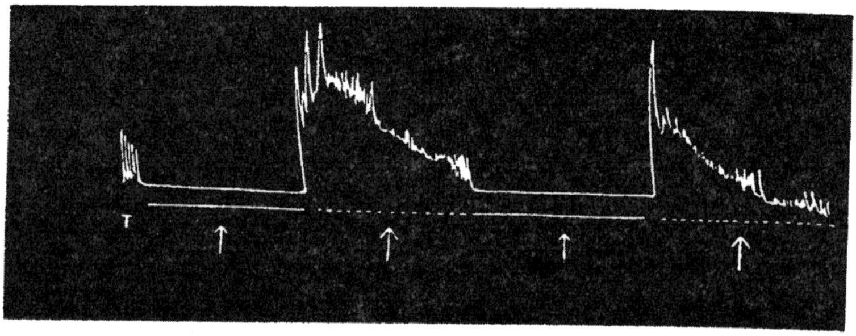

FIG. 69. Direct and after-effect of homodromous current. Transmission of excitation (salt-tetanus T) arrested under homodromous current, denoted by continuous line; on cessation of current, represented by dotted line, there is a transient enhancement above the normal (nerve of Frog).

therefore due entirely to variations of conductivity. The parallel phenomena observed in the transmission of excitation in the animal nerve must, therefore, also be due to induced changes of conductivity.

The action of an electric current in inducing variation of conductivity may be enunciated under the following laws, which are equally applicable to the conducting tissue of the plant and to the nerve of the animal :—

LAWS OF VARIATION OF NERVOUS CONDUCTION UNDER THE ACTION OF A CONSTANT CURRENT

(1) THE PASSAGE OF A CURRENT INDUCES A VARIATION OF CONDUCTIVITY, THE EFFECT DEPENDING ON THE DIRECTION AND INTENSITY OF THE CURRENT.

(2) WHEN IT IS OF FEEBLE INTENSITY, A HETERODROMOUS CURRENT ENCHANCES, AND A HOMODROMOUS CURRENT DEPRESSES, THE CONDUCTION OF EXCITATION.

(3) THE AFTER-EFFECT OF A FEEBLE CURRENT IS A TRANSIENT CONDUCTIVITY-VARIATION WHICH IS OPPOSITE IN SIGN TO THAT INDUCED DURING THE MAINTENANCE OF THE CURRENT.

SUMMARY

The variation of conductivity induced by the directive action of a constant current has been investigated by two different methods:

(1) The method in which the normal speed and its induced variation are automatically recorded;

(2) That in which the variation in intensity of the transmitted excitation is gauged by the varying amplitude of the resulting response.

The effects of the direction and intensity of a constant electric current on the transmission of excitation through the conducting tissue of the plant, may be summarised as follows:—

The velocity of transmission is enhanced when against the direction of a feeble current, and retarded when in the direction of the current. Feeble heterodromous current enhances conductivity, homodromous current depresses it.

Ineffectively transmitted excitation becomes effectively transmitted under a heterodromous current. Effectively transmitted excitation, on the other hand, becomes ineffectively transmitted under the action of a homodromous current.

The after-effect of a current is a transient conductivity-change, the sign of which is opposite to that induced during the passage of the current. The after-effect of a heterodromous current is thus a transient depression, that of homodromous current a transient enhancement, of conductivity.

The characteristic variations of conductivity induced in animal nerve by the direction and intensity of a constant current are in every way similar to those induced in the conducting tissue of the plant.

The above characteristic effects undergo reversal under an intensity of current above a certain critical value.

CHAPTER XIII

COMPLEXITY OF THE MOTILE ORGAN

THE effect of transmitted impulse is, in plants with motile leaves, manifested by the mechanical response of the pulvinus. It is therefore essential to obtain a clear idea of the anatomical and physiological characteristics of the pulvinus, which is not a simple but a highly complex organ. I confine myself here to a brief account of the pulvinus of *Mimosa pudica*: the subject will be treated more comprehensively in a subsequent work.

ANATOMY OF PETIOLE-PULVINAR JUNCTION

A longitudinal section of petiole and pulvinus, reproduced from a photomicrograph, is given in fig. 70, in which the following anatomical features will be noted. The two main bundles, upper and lower (as also two lateral bundles not shown in the figure), while distinct in the petiole, converge and meet in the pulvinus; the pith becomes thereby reduced, and the cortical parenchyma increased proportionately. A very noticeable fact is that though the cylinder of protective sclerenchyma, which extends throughout the whole length of the petiole and the pulvinus, is lignified in the petiole, the lignification of the tissue ceases in the pulvinus, as shown by the usual micro-chemical tests. The significance of the non-lignification of the sclerenchyma-cylinder in the pulvinus is obviously to allow rapid movement of the organ in response to stimulation.

The exact line of demarcation between the petiole and

the pulvinus cannot be ascertained even by microscopic examination; for the cortical cells of the petiole pass imperceptibly into those of the pulvinus. I have, however, discovered a method, by which the line of demarcation can be brought out with extraordinary clearness.[1]

The phloem has been shown to be the conductor of excitation and thus to function as a nerve; in each bundle there are two phloems, one external, the other internal. These nerve-strands of the four bundles in the petiole meet in the pulvinus and form an almost continuous ring; it will be shown later that their nerve-ends in the pulvinus remain functionally distinct.

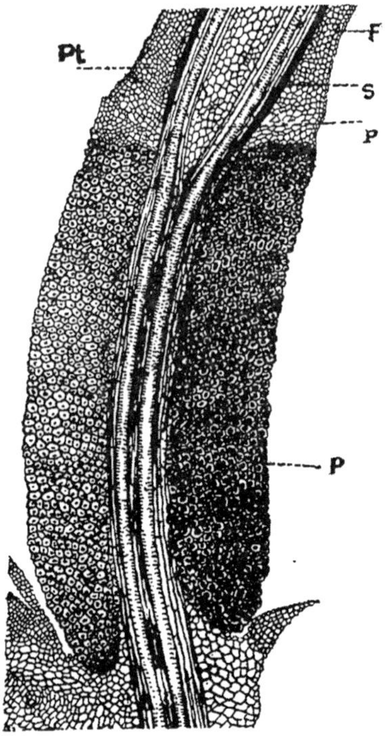

FIG. 70. Longitudinal section of petiole and pulvinus passing through upper and lower vascular bundles.

F, the vascular bundle which meets its fellows in the pulvinus; S, protective sclerenchyma-cylinder, lignification of which extends only through the petiole; P (upper), pith disappearing in pulvinus; Pt, petiole, the cells of which remain unstained; P, contractile cells of pulvinus deeply stained. (*Mimosa pudica.*)

FOUR EFFECTORS IN THE PULVINUS

It is usual to regard the pulvinus as consisting of only two functional halves, an upper and a lower, the responsive movements being rectilinear, up or down. In reality the organ is more complex; for it gives certain other characteristically definite

[1] 'Physiological and Anatomical Investigations on the Conducting and Motor Tissue of *Mimosa pudica*,' *Proc. Roy. Soc.*, B, vol. 98, Aug. 1925.

responses. In addition to the down- and up-movements, the pulvinus exhibits right and left twists, that is to say, torsional clockwise and anti-clockwise responses, depending on which particular flank of the organ has been subjected to stimulation. The motor organ thus consists of four effectors for the production of four distinct types of response.

THE TORSIONAL RECORDER

The record of torsional response is obtained as follows. In order to eliminate the effect of the weight of the leaf

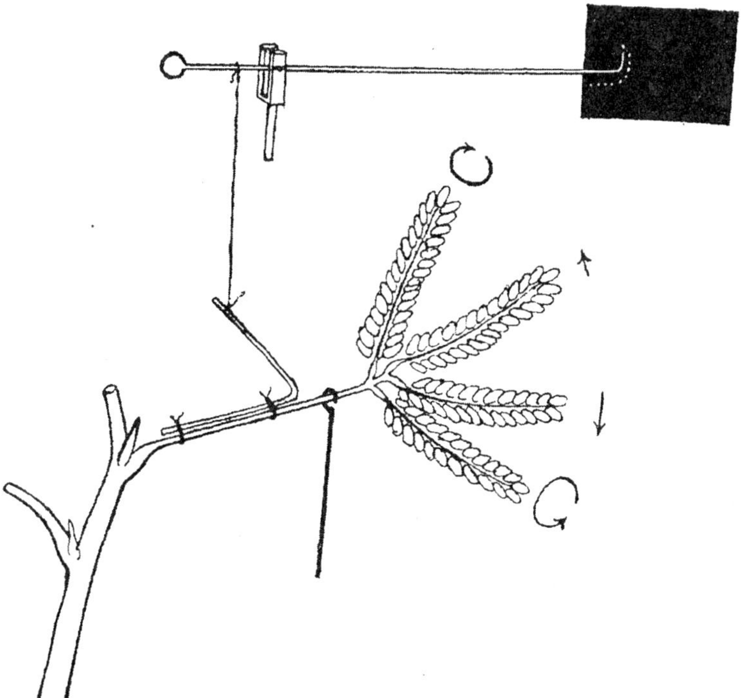

FIG. 71. The Torsional Recorder.
Stimulation of left flank of pulvinus induces an anti-clockwise, and of the right flank a clockwise, torsion.

and to obtain pure torsion, the petiole is held in a hooked glass support with a smooth internal surface, which prevents any up or down movement, yet allowing freedom for torsional response. The torsion is magnified by an L-shaped piece of aluminium wire so tied to the petiole that its free arm is at right angles to the petiole. The end of the arm is connected by a silk thread to the short arm of a recording lever, thus ensuring a compound magnification of the torsional movement : a left-handed torsion produces an up-curve in the record, and a right-handed torsion produces a down-curve. The record is taken by the Oscillating Recorder, the successive dots being at definite intervals of time, which could be varied according to requirements. The same apparatus may be used for obtaining the up- and down-records, when the hooked support is removed and the short arm of the lever directly attached by a thread to the petiole (fig. 71).

The Quadrants of the Pulvinus

The quadrants or effectors are represented diagrammatically in fig. 72. The observer standing in front of

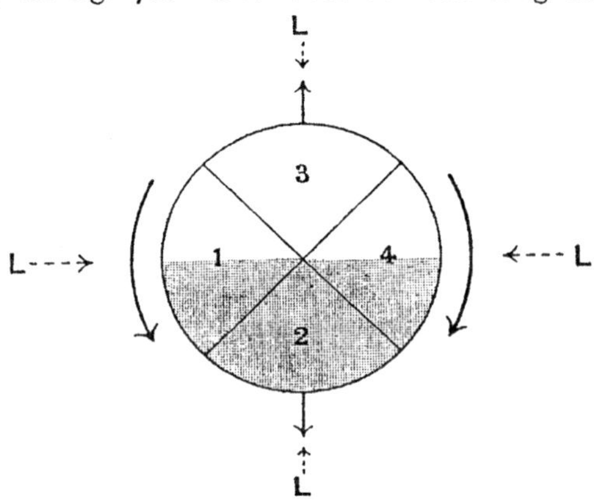

FIG. 72. Diagrammatic representation of the quadrants and their characteristic responses.
Dotted arrow represents direction of incident light, and full arrow the responsive movement. Lower half of pulvinus shaded.

the leaf is supposed to be looking at the stem. The two quadrants to the left and to the right are numbered (1) and (4), respectively: the lower quadrant is numbered (2), the upper (3).

When the stimulus of light acts on the upper quadrant (3) the response is a rectilinear up-movement; when light acts on the lower quadrant (2) from below, the response is a more energetic down-movement: stimulation of quadrant (1) or (4) induces a torsional response.

Lateral Stimulation by Light

Experiment 75. *Effect of stimulation of left quadrant.*—When a beam of light, thrown laterally on the pulvinus, strikes the quadrant (1), the response is a torsional movement which is anti-clockwise, shown as an up-curve (fig. 73). There is recovery after the cessation of exposure to light.

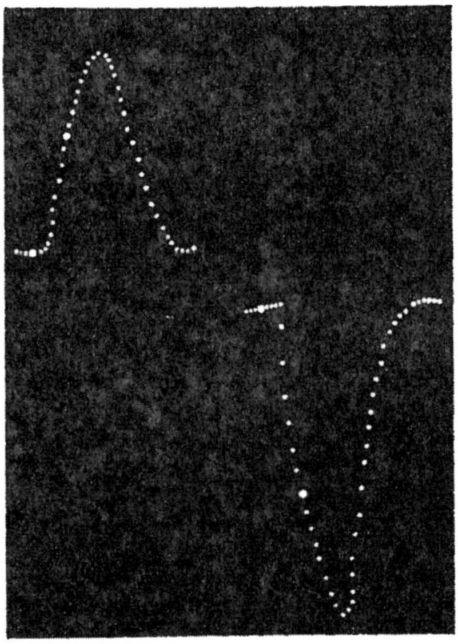

FIG. 73. Records of torsional response due to stimulation of left and right quadrants.

Experiment 76. *Effect of stimulation of right quadrant.*—If the direction of stimulus be changed, so that the light strikes the right quadrant (4), the response is a clockwise torsion, shown as a down-curve (lower record, fig. 73). The intensity of light happened to be stronger in this case, and the amplitude of response was therefore relatively greater.

LATERAL STIMULATION BY OTHER AGENTS

Experiment 77. *Effect of thermal radiation.*—The source of thermal radiation was a length of electrically-heated platinum wire. Stimulation of the left flank induced an anti-clockwise torsion; of the right flank induced a clockwise torsion.

Experiment 78. *Geotropic stimulation.*—By means of suitable inclination, the left flank (1) of the pulvinus was exposed to the action of the vertical lines of force of gravity. The response was an anti-clock or left-handed torsion: when the opposite flank of the organ was similarly stimulated, the response was a right-handed torsion.

Inspection of fig. 72 makes it clear that in all cases the responsive torsion is such that it is the less excitable upper half of the organ that is made to face the stimulus. Similar torsional response is found to be given not only by pulvinated but by all anisotropic organs.

The Law of Torsional Response is:

> AN ANISOTROPIC ORGAN, LATERALLY EXCITED BY ANY STIMULUS, UNDERGOES A TORSION SUCH THAT ITS LESS EXCITABLE SIDE IS MADE TO FACE THE STIMULUS.

THE TORSIONAL BALANCE

A beam of light falling on the left flank of the pulvinus of Mimosa induces an anti-clockwise torsion. If a second beam fall on the right flank, it will induce a clockwise torsion. The two torsions being opposed, the resultant effect is determined by the effective stimulation of the two flanks. The pulvinus may thus serve as a delicate indicator by which the effective intensity of two stimulations may be compared with each other.

Experiment 79. *Comparison of phototropic reactions to light of different colours.*—A parallel beam of light from a small arc-lamp passing through a blue glass falls on the left flank of the pulvinus; a beam of blue light is also thrown upon the opposite right flank, the intensity of the latter being so adjusted that the resultant torsion is zero. The blue glass on the left side is then removed, the unobstructed white light being allowed to fall on the left flank. This causes an upset of the balance, the resultant torsion being anti-clockwise. This is due to the more effective stimulation by the unobstructed white light. A red glass is next interposed on the left side, with the result that balance is upset, this time in the opposite direction. This proves that the phototropic effect of blue light is greater than that of red light.

In this way it is possible to compare the tropic effect of one mode of stimulation with that of a totally different mode, phototropic against geotropic stimulation for example. It is enough here to draw attention to various investigations rendered possible by this very sensitive method of Torsional Balance.[1]

Complex Lateral and Torsional Movements under Vertical Light

Very complicated movements are executed by leaflets of various plants under the action of vertical light. Thus the leaflets of *Cassia alata* close *laterally* in darkness. Under moderate diffuse illumination they open out in a lateral direction. But under strong vertical light, the pulvinules of the leaflets exhibit a torsion by which the formerly infolded surfaces of the leaflets are exposed at right angles to the light from above (fig. 74). Such complicated movements, in two directions of space, are also exhibited by other leaflets which close at night in a lateral direction.

In order to obtain an explanation of these complex movements under different intensities of light, it is necessary to discover the characteristic different excitabilities of the two halves of the pulvinule.

Determination of differential excitabilities of the pulvinule.—In the leaflet of Cassia the movement of opening under

[1] Cf. *Life Movements in Plants*, p. 505 *et seq.*

the stimulus of diffuse light can only be brought about by the contraction of the outer half, which must therefore be the more excitable. This is independently demonstrated by its reaction to an electric shock. On subjecting the half-closed leaflets to diffuse electric stimulation, they open outwards in a *lateral* direction. The relative position of the unequally excitable halves of the pulvinule of Cassia is thus different from that of the main pulvinus of Mimosa. In the latter, the plane that divides the two halves is horizontal, the lower half being the more excitable: whereas in the pulvinule of Cassia, the plane that separates the two

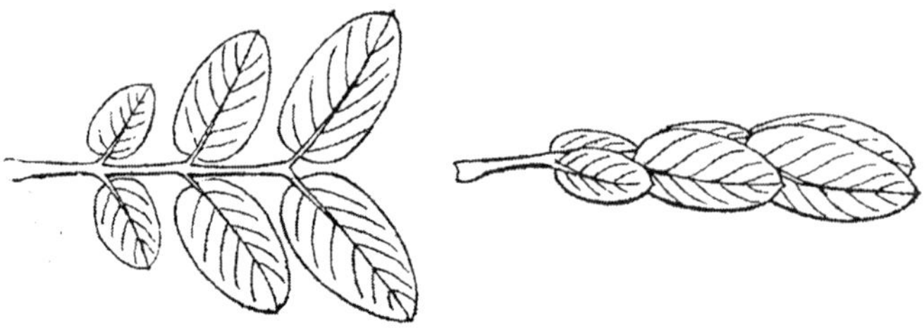

FIG. 74. Leaflets of *Cassia alata*: open in daytime, and closed in evening.

unequally excitable halves is vertical, the outer half being the more excitable. By 'inner half' is here meant that half which is inside when the leaflets are closed.

Effect of strong vertical light.—When the plant is placed in a moderately lighted room, the leaflets open out laterally to the utmost. This is brought about by the contraction of the more excitable outer half of the pulvinule. If strong light be thrown from above, a new movement is superposed, namely, a torsion by which the leaflets undergo a twist so as to place their inner surface at right angles to the vertical light. In order to investigate this phenomenon in greater detail I placed the plant in a well-lighted room, the leaflets being three-quarters open under diffuse light. A very light index was attached to the leaflet for magnifying the subsequent torsional movement. A strong beam of parallel light from an arc-lamp was thrown down on the pulvinule from above; this fell on the junction of the more excitable outer with the less excitable inner half of the organ, the plane of separation of the two unequally excitable halves being, as previously explained, vertical. The leaflet became twisted so as to expose the former infolded surface upwards, at right angles to the incident light. This is further evidence that, under lateral stimulation, a differentially excitable organ undergoes torsion by which the less excitable half is made to face the stimulus.

As a confirmatory test, strong light was made to strike the pulvinule from *below*, with the result that the leaflet exhibited an opposite torsion by which the infolded surface faced downwards, so as to be at right angles to the light incident from below.

Under natural conditions sunlight falls from above; stimulation thus takes place at the junction of the two differentially excitable halves of the organ, the plane of separation of which is vertical. The torsion induced makes the less excitable inner half turn in such a way that the infolded surfaces of the leaflets are placed perpendicular to the incident light.

SUMMARY

The pulvinus of Mimosa has been found to consist physiologically of four different effectors characterised by definite responsive movements of the leaf. The upper and lower quadrants respond to direct stimulation by rectilinear up and down movements, respectively: the left quadrant responds by an anti-clockwise torsion; the response of the right quadrant is a clockwise torsion.

The law of torsional response is, that an anisotropic organ laterally excited by any stimulus, undergoes torsion such that its less excitable side is made to face the stimulus.

The Torsional Balance permits of simultaneous comparison of the tropic effects of two different forms of stimulation.

CHAPTER XIV

THE DIA-HELIOTROPIC ATTITUDE OF LEAVES

THE leaves of plants adjust themselves in various ways in relation to the incident light. The heliotropic fixed position is assumed by means of curvatures and torsions of the motile organ, which may be the pulvinus, or the petiole acting as a diffuse pulvinoid. In some cases the motile organ is both perceptive and responsive; in others, the leaf-blade exerts a directive action, the perceptive lamina and the motor organ being separated by an intervening distance. This directive action of the lamina has been observed by Vöchting in *Malva verticillata*, and by Haberlandt in *Begonia discolor* and in several other plants. It should be borne in mind that this characteristic reaction does not preclude the possibility of the motile organ being directly affected by the stimulus. In a nerve-and-muscle preparation, the muscle is excited not merely by indirect but also by direct stimulation. In the heliotropic adjustment of leaves in the cases just mentioned, the stimulus of light acts both directly and indirectly, the indirect stimulation being due to some impulse transmitted from the perceptive lamina. This may be generally expressed by the statement that the coarse adjustment of the leaf is brought about by direct, and the finer adjustment by indirect, stimulation.

Most leaves assume a heliotropic fixed position such that the blades are placed at right angles to the incident light with the ventral surface uppermost, the directive action being due to some as yet unknown transmitted impulse. Nor has any satisfactory explanation been forthcoming of the physiological reaction of which this movement is the expression. Suggestions have been made that the dia-heliotropic position of leaves is of obvious advantage, since it assures for the plant the maximum illumination. But such teleological considerations offer no explanation of the definite physiological reaction. It is, moreover, not true, as I show in the course of this chapter, that there is anything inherent in the irritability of the plant which constrains the surface of the leaf to place itself perpendicular to the incident light.

The dia-heliotropic phenomena have been studied not only in 'sensitive' but also in ordinary plants: it will be shown that the responsive reactions are essentially similar in both. As a type of the former I have taken *Mimosa pudica*, and of the latter, *Helianthus annuus*.

General Description of Dia-Heliotropic Phenomena

Before entering upon the experimental investigation of the subject, I will describe the dia-heliotropic phenomena, as manifested by Mimosa and Helianthus. A photograph of the former is reproduced in fig. 75, a. The plant, grown in a pot, had been exposed to the northern sky and not to direct sunlight. It will be seen that the leaves which directly front the light have been raised, and so placed that the sub-petioles, with their leaflets, are at right angles to the strongest illumination. The lateral leaves have, on the other hand, undergone appropriate torsion, the plane of the leaflets being adjusted perpendicular to the light: it will be noticed that the petioles to the right and the left have undergone opposite torsions. After the assumption of this position, the pot containing the plant was turned round through 180°. This brought about a new adjustment in the course of twenty minutes, the plane of all the leaflets being once more at right angles to the light. The new adjustment necessitated a complete reversal of the former movements and torsions. Such perfect adjustment is best brought about by bright light

FIG. 75.—Dia-heliotropic adjustment of leaves.
(a) in *Mimosa pudica*; (b) in *Helianthus annuus*; and (c) in another species of Helianthus.
(From photographs.)

from the sky, not so well by direct sunlight, for reasons which will be given later.

Figure 75, b shows the heliotropic adjustment of the leaves of a Sunflower, grown near a wall, the plant being exposed to light from the western sky. The adjustment is essentially similar to that seen in Mimosa. The lateral leaves, (1) and (3), have undergone appropriate torsion—right-handed or left-handed—such that the leaf-blades are placed at right angles to the light with the ventral surface always facing it. The leaf number (2) has been raised, placing its lamina perpendicular to the light. A contributing factor is the bending over of the stem, due to positive heliotropic curvature, which accentuates the rise of the leaf number (2). The same bending often causes an apparent fall of the leaf marked (4). When the stem is tied to a stake, the bending over of the stem is prevented; the leaf numbered (2) is then raised by its own heliotropic reaction; but there is little or no fall of the opposite leaf.

Another photograph is reproduced (fig. 75, c) of the heliotropic curvature and adjustment of a different species of Sunflower, which was grown in the open. In the morning the plant bent over to the east and all the leaves exhibited appropriate dia-heliotropic movements and torsions. In the afternoon the plant bent over to the west, all the previous adjustments and torsions being completely reversed. The plant continued to exhibit these alternate swings day after day till the movement ceased with age.

Mechanical Response due to Differential Excitability

It has already been explained that on account of the differential excitability of the upper and lower halves of the pulvinus, a diffuse stimulus causes a responsive fall of the leaf of Mimosa. I shall now show that a similar reaction is manifested by Helianthus.

Experiment 80.—In Helianthus, the entire petiole acts as a motile organ, of which the upper half is relatively the less excitable. Diffuse stimulation by electric shock induces a responsive fall, followed by a recovery on the cessation of stimulation. The response-records thus obtained are very similar to those obtained with the leaf of Mimosa. In Helianthus the reaction is relatively sluggish and the contraction is not so great as in Mimosa. The difference between the two responsive reactions is, however, merely one of degree and not of kind.

Response to Stimulation of Upper and Lower Halves of the Organ

The pulvinus of Mimosa responds to the application of light from above by contraction of its upper or ventral and expansion of the lower or dorsal half; the leaf is thus erected, and the movement towards the light may be described as positive heliotropism. The leaflets attached to the sub-petioles are thus made to face the light. Under strong and long-continued sunlight the excitation is transmitted across the pulvinus, causing at first a neutralisation, and finally a reversed or negative movement due to the contraction of the more excitable lower half of the organ. This is the reason why the dia-heliotropic adjustment is less perfect under strong sunlight than under diffuse light.

Experiment 81.—A parallel reaction is obtained with Helianthus; here the petiole acts as an extended pulvinoid. Light applied from above causes an erectile movement; when applied below it causes a more energetic down-movement. As the transverse conductivity of the petiole is feeble, the positive heliotropic response, induced by light acting from above, is rarely reversed into negative.

The Mechanism of Heliotropic Curvature

A few words may now be said on the mechanics of the curvature by which the stem of Helianthus bends towards light. All forms of stimulation, including that of light, induce a diminution of turgor and consequent contraction, and retardation of the rate of growth of the directly excited side. But this is not the only factor in bringing about the positive curvature. I have shown that while the effect of direct stimulation of the proximal side of the stem is to induce a diminution of turgor and contraction, its effect on the distal side, where it acts indirectly, is the very opposite, namely, an increase of turgor and expansion. The positive curvature is thus due to the joint effect of direct stimulation of the one side and indirect stimulation of the other. I have already demonstrated the induced increase of turgor at the distal side in an experiment with the stem of Mimosa (p. 88). The stimulus of light was applied at a point directly opposite to the motile leaf, which by its movement indicates the change of turgor, the induced increase of turgor being indicated by an erection, and diminution of turgor by a fall of the leaf. Application of light at a point on one side of the stem was thus found to induce an increase of turgor at its diametrically opposite point, as evidenced by the erectile movement of the leaf. (See fig. 33.)

Receptor, Conductor and Effector

The nervous system of plants must be regarded as of a comparatively simple type. In connexion with the evolution of the animal nervous system, it has been suggested that the contractile tissue or muscle appeared first as an independent effector, and that the nerve developed secondarily in conjunction with the muscles as a means of quickly setting them in action; that a receptor or sense-organ alone would be of no service to an organism, neither would nerve or nerve-centres alone; whereas a muscle-cell or effector is of use if it can be stimulated directly.

In plants there are clear indications of these different stages. Thus in the leaf of *Erythrina indica*, and in the terminal leaflet of *Desmodium gyrans*, the pulvinus is the independent effector: heliotropic movement only takes place when the pulvinus is directly stimulated, illumination of the lamina having no effect; the connecting nerve-link is absent or functionally ineffective. In Mimosa and in Helianthus, on the other hand, the intermediate nerve-network is effective, the leaflets or the lamina serving as receptive organs. Haberlandt has shown that in many cases the epidermal cells of leaves are of a lenticular shape, for increasing the perception of light. He rightly observes that 'in zoological nomenclature, organs concerned with the perception of external stimuli have always been known as sense-organs, even among lower animals and in other cases in which it is doubtful if the organs in question are responsible for sensation in the psychological sense. It is, therefore, not only permissible, but necessary in the interest of consistency to apply the term sense-organ to the analogous structure in plants.' [1]

Since the nervous reactions in animals and plants are so essentially similar, delay in full recognition of this fact will undoubtedly retard the advance of science. I shall in the present chapter demonstrate certain striking effects in plants, which at first sight may no doubt appear to be very surprising, but which can be simply explained as the result of nervous reactions usually regarded as the special characteristic of the animal. I will produce evidence that in the plant a definite nervous link exists between the receptor and the effector, and that there is a well-developed system of innervation, by which the 'attitude' of the plant-organ becomes adjusted to the incident stimulus.

DEFINITE INNERVATION OF THE MOTILE ORGAN

I have explained in the previous chapter that the pulvinus of Mimosa is a highly complex organ, its four different quadrants functioning as so many different effectors. When the upper quadrant is directly stimulated by light, there is a responsive up-movement of the leaf; when the lower quadrant is stimulated, the response is a down-movement. Stimulation of the left quadrant induces

[1] Haberlandt, G.—*Physiological Plant Anatomy*—English translation, 1914, p. 572.

a left-handed torsion; that of the right quadrant, a right-handed torsion.

The characteristic reaction of each of the quadrants to direct stimulation can also be obtained by the indirect stimulation of the corresponding sub-petiole: this is evidence that a definite nerve-connection exists between each of the sub-petioles and the corresponding quadrant of the pulvinus. The characteristic responses are the same whatever the mode of stimulation, whether mechanical, electric, or photic. Care has, however, to be taken that the stimulus is not too strong, for the effect of excessive stimulation becomes diffused and thus causes a fall of the leaf by the predominant contraction of the lower half of the pulvinus. Electric stimulation by tetanising electric shocks has the advantage that the intensity may be reduced to any extent desirable. Photic stimulation may be effected by throwing light from an arc-lamp on a particular sub-petiole, the intensity of stimulation being suitably increased by prolonging the exposure. The stimulation effected by light is less intense than that caused by electric shocks, and the reaction under light is, therefore, relatively sluggish. The several sub-petioles will be distinguished by definite numbers, counting from the left, the observer being supposed to face the central stem. The sub-petioles (1) and (4) are the two extremes, the two intermediate ones being (2) and (3).

Characteristic Leaf-Movements on Stimulation of the Several Sub-Petioles

Experiment 82. *Response to electric stimulation of sub-petioles* (1) *and* (4).—Sub-petiole number (1) was first stimulated by a moderately feeble electric stimulus of short duration; the response was a *left-handed* torsion of the leaf, represented by an up-curve. The response was initiated in a short time, and there was recovery on the cessation of stimulation. Sub-petiole number (4) was next subjected to stimulation and gave a similar torsional response, but now in a *right-handed* direction, recorded as a down-curve (fig. 76, E).

Experiment 83. *Response to photic stimulation of* (1) *and* (4).—In order to demonstrate that different modes of

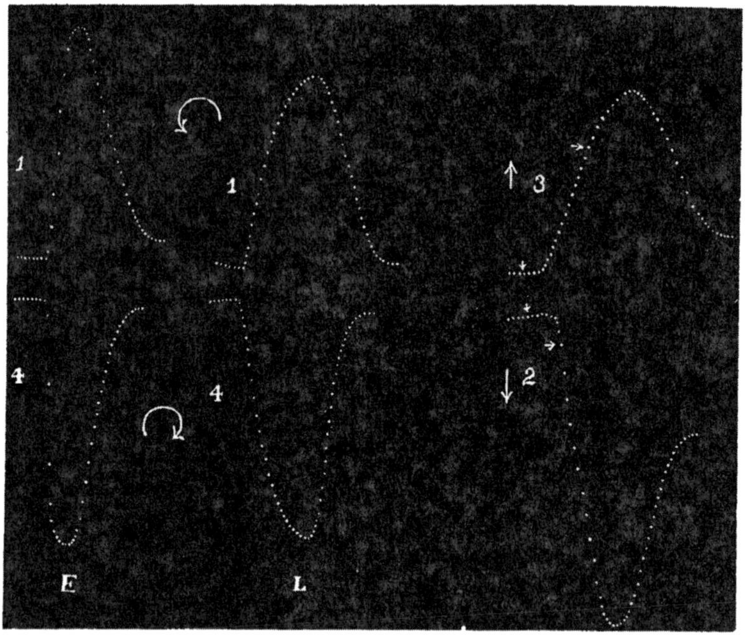

FIG. 76. Record of responses of leaf to stimulation of the several sub-petioles of *Mimosa pudica*.

E, torsional responses under electric stimulus; stimulation of sub-petiole 1 causes left-handed torsion (up-curve); that of 4, a right-handed torsion (down-curve). L, similar torsional responses under stimulus of light. The records to the extreme right show the rectilinear responses induced by stimulation of 2 (rapid down-movement) and of 3 (slow up-movement).

stimulation induce an identical effect, the sub-petiole (1) was stimulated by throwing on it a strong beam of light from an electric lantern. The response was, as in the last case, a left-handed torsion: stimulation of number (4) induced a right-handed torsion (fig. 76, L).

Experiment 84. *Stimulation of sub-petioles* (2) *and* (3).—The responsive movements of the leaf caused by stimulation by light of the two intermediate sub-petioles numbers (2) and (3) were as follows. Stimulation of sub-petiole number (2) caused a rapid responsive fall, followed by subsequent recovery on the cessation of light. Stimulation of sub-petiole (3) gave rise to a relatively slow up-movement; the duration of the exposure to light had, in this case, to be prolonged in order to obtain a moderate amplitude of response.

The effects of direct and indirect stimulation may be summarised as follows:

(1) Direct stimulation of the left quadrant or effector (1) induces an anti-clockwise torsion; indirect stimulation transmitted from sub-petiole (1) produces the same effect.

(2) Direct stimulation of the lower effector (2) induces a rectilinear down-movement; the same effect is produced by indirect stimulation transmitted from sub-petiole (2).

(3) Direct stimulation of the upper effector (3) induces a rectilinear up-movement; the same effect is produced by indirect stimulation transmitted from sub-petiole (3).

(4) Direct stimulation of the right effector (4) induces a clockwise torsion; the same effect is produced by indirect stimulation transmitted from sub-petiole (4).

These results prove that there is a definite nervous connection between each sub-petiole and its corresponding quadrant or effector in the pulvinus, such that peripheral stimulation causes the response characteristic of each particular quadrant when directly stimulated (fig. 77). The nerve-connection between the periphery and centre is also independently demonstrated by experiments described in the next chapter (*cf.* Experiment 85).

Adjustment of Leaf in Space by Transmitted Nervous Excitation

Having established the characteristic innervation of the pulvinus, it is now possible to offer a satisfactory explanation of the dia-heliotropic attitude of the leaf under peripheral stimulation of the sub-petioles by light.

When the leaflets, say, of the right sub-petiole, are acted on by vertical light, the distant pulvinus responds

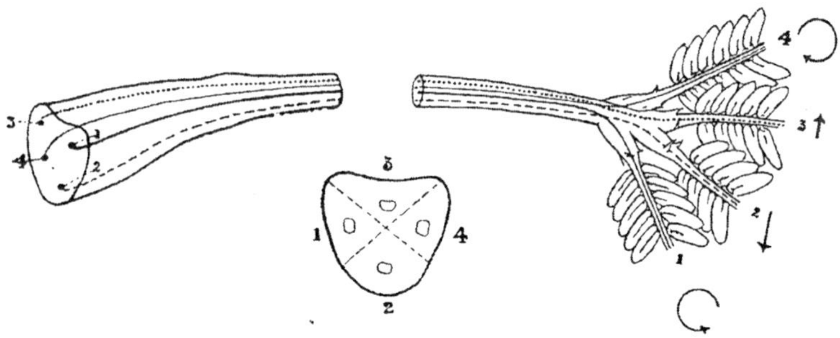

FIG. 77. Showing the course of four nerve-strands from the four sub-petioles to the pulvinus. (*Mimosa pudica.*)

The lower figure is a diagrammatic section of the pulvinus with its four effectors. Effectors 1 and 4, which give rise to left- and right-handed torsions, are respectively in nervous connection with sub-petioles 1 and 4. The lower effector 2 is connected with sub-petiole 2, the response being a rapid down-movement. The upper quadrant 3 is in connection with sub-petiole 3, the response being a slow up-movement.

with a right-handed torsion. The leaflets on the sub-petiole are thus carried passively like so many flags, the amount of light absorbed by them being considerably reduced. It is therefore not the advantage of the plant, but the inevitable physiological reaction, that determines the movement. But when the two sub-petioles (1) and (4) are simultaneously exposed to light of the same intensity, the two resultant torsions balance each other. While in this dynamic balance, if the intensity of light on, say, the left sub-petiole (1), be diminished by interposition of a piece of paper, the balance is at once upset, and there is a resultant right-handed torsion. Hence the lateral adjustments of the leaf as a whole are made by the two sub-petioles (1) and (4) which are situated externally.

The balancing adjustments, up or down, are made in response to excitation transmitted by the two middle sub-petioles. It is clear that equilibrium is only possible when the entire leaf-surface (consisting of the leaflets carried by the four sub-petioles) is equally illuminated; and this can only occur when the leaf-surface as a whole is perpendicular to the incident light. The leaf is adjusted in space by the co-ordinated action of four reflexes. The dia-heliotropic attitude of leaves is thus the resultant effect of distinct nervous impulses initiated at the perceptive region and actuating the different effectors at a distance.[1]

SUMMARY

It has been shown that in certain leaves the heliotropic adjustment is brought about by the transmission of nervous impulses from the perceptive to the motor organ. Continuity is shown to exist between the response of 'sensitive' and that of ordinary plants, *Mimosa pudica* being the type of the former, and *Helianthus annuus* of the latter. Mechanical response is brought about in both by the differential excitability of the upper and the lower halves of the motile organ. The lower half in both is the more excitable. Local stimulation of the upper half of the organ induces an erectile movement: that of the lower half, a more rapid downward movement.

Heliotropic curvature of the stem is due to the joint effects of contractile reaction of the proximal and of expansive reaction of the distal side.

The pulvinus of *Mimosa pudica* may be regarded as consisting of four effectors; under direct stimulation the

[1] For a more detailed account *cf.* 'The Dia-Heliotropic Attitude of Leaves as Determined by Transmitted Nervous Excitation,' *Proc. Roy. Soc.*, B, vol. 93, 1922, p. 153.

response of the left effector is by an anti-clockwise torsion, and of the right effector by a clockwise torsion. The upper and lower effectors respond by rectilinear up-and-down movements.

The characteristic responses described above are also given under indirect stimulation. The four quadrants respond to excitation transmitted from the four corresponding sub-petioles when stimulated, the resulting response being determined by the characteristic reaction of the particular effector.

Stimulation of the leaflets of sub-petiole (1) by light gives rise to an excitatory impulse which, on reaching the left effector, induces a left-handed torsion; stimulation of those of sub-petiole (4) induces the opposite or right-handed torsion. Illumination of the leaflets of sub-petiole (2) induces a down-movement, illumination of those of sub-petiole (3) an up-movement. The leaf is thus adjusted in space by the co-ordinated action of four reflexes, equilibrium being attained when the leaf-surface is perpendicular to the incident light.

The dia-heliotropic attitude of the leaf, as exemplified by *Mimosa pudica*, is brought about by the characteristic reactions of the several effectors in response to transmitted nervous impulses initiated at the perceptive region of the lamina.

CHAPTER XV

THE REFLEX ARC IN MIMOSA AND THE TRANSFORMATION OF AFFERENT INTO EFFERENT IMPULSE

THE definite innervation of the four quadrants of the pulvinus of *Mimosa pudica* was demonstrated in the last chapter. It was shown that the nerve from the sub-petiole (1) is led to the left quadrant; the sub-petiole (2) is in nervous connection with the lower quadrant; the central nerve-termination of sub-petiole (3) is in the upper quadrant; and finally the nerve of sub-petiole (4) terminates in the right quadrant.

NERVE-CONNECTION BETWEEN CENTRE AND PERIPHERY

I have succeeded in further demonstrating by an independent method the nerve-connection between each quadrant and the corresponding sub-petiole. It consists in observing the effect of stimulating each of the four nerve-ends imbedded in the four quadrants of the pulvinus. In Experiment 19, it was shown that *superficial* 'scratch-stimulus' applied to the sensitive cortex caused contraction and fall of the leaf without exciting the enclosed nerve-ends. These can be separately stimulated by thrusting a sharp-pointed pin into each of the four quadrants till the nerve-end is reached. After a little practice, it is easy to feel the moment when the pin touches the vascular bundle containing the nerve. It has been explained that the vascular bundles at the centre coalesce into a continuous ring in which the nerves nevertheless remain functionally distinct.

Experiment 85.—When the nerve-end in the left quadrant (1) is cautiously stimulated, an excitatory impulse is generated which travelling outwards causes closure only of the leaflets on sub-petiole (1). Stimulation of the nerve in the lower quadrant (2) causes the same response in sub-petiole (2). Similarly stimulation of the nerve-ends in quadrants (3) and (4) elicits corresponding response in sub-petioles (3) and (4) respectively. In the following table is given a synopsis of forty different experiments, ten experiments on the stimulation of each nerve-ending. The transmission-time was found by measuring the interval between the application of stimulus and the closure of the innermost pair of leaflets on each sub-petiole. The distance traversed was the length of the petiole *plus* a very short length of the sub-petiole. The length of the petiole was not exactly the same in the different cases; the transmission-time is the average of ten different experiments.

TABLE VIII.—TRANSMISSION-TIME OF IMPULSE FROM CENTRE TO PERIPHERY

Stimulation at central end	No. of experiments	Response of	Transmission-time
Left quadrant nerve (1)	10	1st sub-petiole	25 sec.
Lower ,, ,, (2)	10	2nd sub-petiole	24 sec.
Upper ,, ,, (3)	10	3rd sub-petiole	24 sec.
Right ,, ,, (4)	10	4th sub-petiole	23 sec.

The average transmission-time given in the above table is 24 seconds. In a second series of experiments the average value was found to be 22 seconds. The average transmission-time of the centrifugal impulse from the centre to the first pair of leaflets in the sub-petiole is therefore about 23 seconds more or less.

Conversely it is possible to determine the transmission-time of the centripetal impulse in the petiole initiated by peripheral stimulation. A moderately strong electric stimulus is applied at the middle of one or other of the sub-petioles. The excitatory impulse travels towards the centre. On account of the intensity of the transmitted excitation, it does not remain localised at one or other of the quadrants, but becomes diffused, causing the fall of the leaf. The transmission-time is measured by noting the interval between the closure of the innermost pair of leaflets on the stimulated sub-petiole and the subsequent fall of the leaf.

A detailed account of the results is given in Table IX (p. 182); the average value of the centripetal transmission-time through the petiole, *plus* the short length of the sub-petiole, is 21 seconds. Taking into account the different mode of stimulation and the different batches of plants, the agreement between the transmission-time of this and that of the last series of experiments is remarkably close.

Reflection of Afferent as Efferent Impulse

It has been shown that a minimal stimulus applied to one of the four sub-petioles gives rise to an 'ingoing' impulse which evokes the response characteristic of the corresponding quadrant of the pulvinus. The impulse does not irradiate any further but remains localised.

When the intensity of the stimulus is increased, a new and significant class of phenomena makes its appearance, now to be described.

Fundamental experiment 86.—A short length of the middle of sub-petiole (1) was stimulated for one second by a tetanising induction-current of moderate intensity. The excitatory impulse, travelling through the length of the petiole in an ingoing or centripetal direction, reached the pulvinus, and caused the fall of the leaf. A rod padded with soft cotton-wool had been placed 5 mm. below the petiole to break the fall. *After the arrival of the ingoing or afferent impulse at the pulvinus, it became reflected as an outgoing or efferent impulse,* reaching the sub-petiole (2) with resulting serial fall of its leaflets. It has been shown that an ingoing impulse caused by stimulation of sub-petiole (1) reaches the central end in the pulvinus along nerve (1). It has been shown further (Experiment 85) that stimulation of the central end of nerve (2) gives rise to an outgoing impulse which causes response in sub-petiole (2). In the present case the central end of nerve (2) was not *directly* stimulated; it could only have been excited *indirectly* by the impulse which reached the central end of nerve (1) as the result of the peripheral stimulation of sub-petiole (1). The afferent impulse must then have overflowed or irradiated from the central end of nerve (1) to that of (2), thus forming an arc; *the reflected impulse now travelled along a new path in an outgoing direction.* This is the REFLEX ARC in Mimosa. The ingoing impulse from the peripheral receptor may be variously described as centripetal, afferent or 'sensory': the reflected outgoing impulse may be designated as centrifugal, efferent or 'motor.' The significance of the words in italics will shortly become apparent.

In the experiment described, the transmission-time of the afferent impulse from the innermost pair of leaflets of sub-petiole (1) to the pulvinus was 18 seconds, while the transmission-time of the efferent impulse through practically the same distance was only 3 seconds. *The velocity of the efferent was found in all cases to be greater than that of the afferent impulse.* The relatively high velocity of conduction in the efferent direction is the more remarkable when the total distance which the afferent and efferent impulses had to travel is taken into account: it would naturally be imagined that an impulse would be increasingly impeded the greater the distance it had to travel. But while the primary afferent impulse travelled somewhat slowly in the ingoing direction, after reflection into a new path its velocity was enhanced at least six times.

The essential similarity of nervous impulse in plants and animals has already been demonstrated by the foregoing experiments. The study of the reflex arc in Mimosa is therefore of interest as throwing light on the corresponding phenomenon in the animal. Mimosa indeed offers certain unique advantages. For the moment of arrival of the afferent or 'sensory' impulse is easily detected by the mechanical response of the main pulvinus at the central end, which serves for the determination of the velocity of the afferent impulse. The interval between the fall of the leaf and the closure of the innermost pair of leaflets of sub-petiole (2) gives data for estimating the velocity of the efferent impulse. Again, as the velocity of the nervous impulse in Mimosa is about 1000 times slower than that in the animal, any difference in the velocity of afferent and efferent impulses is easily detected in the plant; this is practically impossible in the animal.

Localisation of the Reflex Arc

The nerve-strand of sub-petiole (1) extends along the petiole to the central or pulvinar end. The wave of excitation in nerve (1) might have passed laterally into nerve (2) in the petiole itself, without taking the longer journey involved in reaching the pulvinus and then returning. Is there any evidence that the impulse actually takes the longer path? It is to be found in the fact that, under normal conditions, the reflected impulse is initiated *after* the arrival of the afferent impulse at the pulvinus. There is a definite sequence of events: first the initiation of peripheral excitation by the stimulation of the sub-petiole; then the conduction of the afferent impulse to the pulvinus, signalled by the fall of the leaf; and finally the arrival of the reflected efferent impulse at the periphery. The average time for the double journey up and down the petiole (obtained from a large number of experiments) comes to about 24 seconds. (see Table IX, p. 182): had the afferent impulse crossed over immediately after reaching the petiole, the total time would have been but a few seconds only. The reason why the impulse travels through the longer distance is that, of the two alternative paths, it follows the line of least resistance. The nerve-strands in the petiole are, as previously explained, remote from each other; while in the pulvinus they are brought into the closest contiguity. The advantage of the shorter path across the petiole is more than counterbalanced by the great resistance to be overcome in passing across the area of imperfectly conducting tissue that intervenes between the nerve-strands. Hence, under ordinary conditions, excitation passes from one nerve to another only at the central or pulvinar end.

Irradiation of Excitation

Two alternative theories may be proposed to account for the passage of excitation at the central end from one nerve to another. The transmission may either be regarded as having been effected through the intervention of the excited pulvinus, or be ascribed to nervous overflow or irradiation from the end of one nerve to that of another.

With regard to the first view, since the reflected impulse is normally initiated after the fall of the leaf, it may be thought that the excitation is transmitted from nerve (1) to nerve (2) *via* the excited pulvinus. It has however been shown (Experiment 19) that, while an excitatory impulse can travel from the nerve to the contractile tissue, it cannot proceed in the opposite direction from the contractile tissue to the nerve. The following experiment, showing the persistence of the reflection of impulse even after abolition of contractility in the pulvinus, proves that the reaction of the motor cells has nothing to do with the efferent impulse.

Experiment 87. *Persistence of reflected impulse after abolition of contractility of the motor cells.*—I have explained that the contractile power of the pulvinus is temporarily abolished by the absorption of an excessive quantity of water. When the pulvinus has thus become immotile, an afferent impulse which reaches it cannot bring about the fall of the leaf, but it is nevertheless reflected as an efferent impulse. It is thus clear that there is no causal relation between the contraction of the pulvinus and the reflection of the impulse. The efferent impulse must therefore be due to a reflex arc. The irradiation from one nerve-end to another takes place whether the pulvinus contracts or not.

Relative Sensibility of the Different Motile Organs

Since all the various motile organs of *Mimosa pudica* are employed in the detection of the excitatory impulse, it is necessary to say a few words about their relative sensibility. A very feeble stimulus causes the closure of the leaflets, the latent period of the pulvinules being only a very small fraction of a second: their sensibility is, moreover, not readily affected by adverse conditions. The excitability of the main pulvinus is not so great: it is affected by adverse conditions, such as diminished intensity of daylight, which cause depression of its sensibility. On very damp days during the rains, the turgor of the organ becomes excessive, and it is then relatively insensitive. The depression of sensibility of the pulvinus so induced may cause such a prolongation of its latent period that the leaflets exhibit closure even before the fall of the leaf. All these aberrations are avoided, by choosing a vigorous plant and a bright and warm day for the observation of normal effects. The secondary pulvinus of the sub-petiole is the least sensitive of all. Thus an impulse of moderate intensity which excites the main pulvinus at the centre and the sensitive leaflets at the periphery, fails to cause any movement of the secondary pulvinus which is interposed between the two ends; it is only under stimulus of considerable intensity that this organ exhibits any responsive movement. This relates to the autumn season, when the majority of these experiments were carried out. The excitability of the secondary pulvinus is enhanced in summer, specially on warm and bright days. I will presently indicate the artificial means by which it is possible to increase its sensibility.

Reflected Impulses on Stimulation of the several Sub-petioles

I go on to describe the reflexes produced on stimulation of the second, third and fourth sub-petioles. The characteristic results described below are obtainable under different modes of stimulation. Thermal stimulation may be employed, but this stimulus cannot be maintained constant in successive experiments. For quantitative investigation it is therefore preferable to stimulate by an induction-current. The characteristic effects of stimulation of the different sub-petioles may, of course, be observed by using a fresh specimen in each case; but it is far more interesting to obtain all the characteristic effects with an identical leaf. This is possible when the specimen is in such vigorous condition that it does not exhibit fatigue during the course of a somewhat prolonged investigation. It is also necessary to choose conditions such that light and temperature remain unchanged. On a bright day these remain fairly constant at about noon, when the temperature is 30°–31° C. (August). On a cloudless day the experiments were carried out in a glass-house facing north. On cloudy days the plant was taken inside the laboratory and exposed to light of constant intensity given by an incandescent electric lamp of 500 candle-power.

Experiment 88. *Stimulation by induction-current.*—The intensity of stimulus employed was slightly above the minimal, and was maintained constant for all the experiments in the series. The duration of stimulation was one second. The transmission-time was slightly shortened under a stronger intensity of stimulus, and increased when the petiole was longer. In the present series, the intensity of stimulus was constant, and the length of petiole along which the impulse had to travel was practically the same throughout.

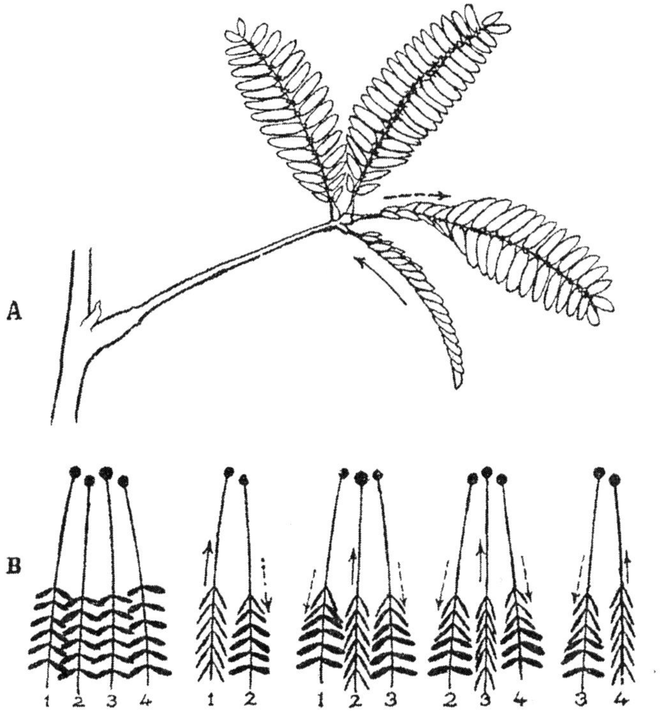

FIG. 78. Reflected impulses initiated by stimulation of the several sub-petioles.

A. Moderate stimulation of sub-petiole (1) gives rise to an efferent impulse which causes fall of leaf (not shown in the figure). The afferent impulse is converted into an efferent after reflection at centre, which causes response of leaflets on sub-petiole (2). Afferent impulse represented by full arrow and efferent by dotted arrow.

B. Diagram illustrating the relation of the four nerve-ends in the pulvinus, and the responses of the leaflets to a single reflected impulse on stimulation of sub-petioles (1) and (4), and to two reflected impulses on stimulation of sub-petioles (2) and (3).

Experiment 89. *Effect of the stimulation of sub-petiole* (1).—Stimulation of sub-petiole (1) gave rise to a single reflected impulse, which caused the closure of the leaflets of only sub-petiole (2). The transmission-time of the afferent impulse was 18 seconds, while that of the efferent was 2 seconds.

Experiment 90. *Effect of the stimulation of sub-petiole* (2).—The afferent impulse reached the pulvinus after 23 seconds, and *the reflected impulse was not single, but double.* These two impulses caused closure of the leaflets of the sub-petioles (1) and (3), 4 seconds after reflection at the pulvinus. The velocity of the efferent was, as in the last case, considerably higher than that of the afferent impulse.

Experiment 91. *Effect of the stimulation of sub-petiole* (3).—The reflected impulse in this case also was *double*; one of the two efferent impulses caused response in sub-petiole (2) and the other in sub-petiole (4). The transmission-time for the afferent impulse was 20 seconds, while that for the efferent was only 3 seconds. The velocity of the out-going impulse was here more than six times greater than that of the in-going.

TABLE IX.—TRANSMISSION-TIMES OF AFFERENT AND EFFERENT IMPULSES

Effect of stimulation by induction current on	Nature of reflected impulse	Transmission-time of afferent impulse	Transmission-time of efferent impulse	Ratio of two velocities
Sub-petiole (1)	Single; response on sub-petiole (2)	18 sec.	2 sec.	1 : 9
Sub-petiole (2)	Double; response on sub-petioles (1) and (3)	23 sec.	4 sec.	1 : 5·8
Sub-petiole (3)	Double; response on sub-petioles (2) and (4)	20 sec.	3 sec.	1 : 6·6
Sub-petiole (4)	Single; response on sub-petiole (3)	24 sec.	3 sec.	1 : 8
	Average	21 sec.	3 sec.	1 : 7

Experiment 92. *Effect of the stimulation of sub-petiole* (4).—The afferent impulse after reflection gave rise

to a *single* efferent impulse as in the case of stimulation of sub-petiole (1). The efferent impulse caused the closure of the leaflets of only sub-petiole (3). The transmission-time for the afferent impulse was 24 seconds, that for the efferent 3 seconds only.

Reference to fig. 78 will facilitate a clearer understanding of the principal results.

Relation between the Afferent and the Efferent Impulses

It is now possible to offer a full explanation of the characteristic effects brought about by peripheral stimulation of the several sub-petioles (*cf.* fig. 78). The ends of the different nerves, as previously explained, are brought into the closest contiguity in the pulvinus. The following relation is thereby established between afferent and efferent impulses in the four sub-petioles :—

(*a*) The nerve-end (1) is in close contiguity to that of (2). The afferent impulse caused by stimulation of sub-petiole (1) is transferred at the centre from nerve-end (1) to nerve-end (2). The efferent impulse therefore causes response of the leaflets on sub-petiole (2).

(*b*) The nerve-end (2) is intermediate between the nerve-ends (1) and (3). The afferent impulse originated in sub-petiole (2) is irradiated into nerves (1) and (3); there are therefore two efferent impulses causing response in sub-petioles (1) and (3).

(*c*) The nerve-end (3) occupies an intermediate position between the nerve-ends (2) and (4). The afferent impulse in (3) therefore gives rise to two reflected impulses which cause response in sub-petioles (2) and (4).

(*d*) Finally the nerve-end (4) to the extreme right is contiguous only to nerve-end (3). The reflected impulse is therefore single; the efferent impulse conducted by nerve (3) causes response of only sub-petiole (3).

The important conclusion established by the experiments described above is the existence of a reflex arc in the pulvinus. Equally significant is the fact that the efferent impulse is far quicker than the afferent. The transmission-time of the latter, through the average length of the petiole (30 mm.) and the short length of sub-petiole (2 mm.), was about 21 seconds. But the mean value of velocity of transmission in the petiole has been found to be about 16 mm. per second (Chapter VI.). Hence the transmission-time (21 seconds) of the afferent impulse would appear to be abnormally high; it should have been about 2 seconds or so, since the distance of transmission is about 32 mm. The anomaly, I shall show, is only apparent and not real. The point is that in the determination of the normal velocity of transmission in the petiole, the stimulus was applied at a certain distance from the pulvinus on the *petiole itself*. But in the determination of the transmission-time of the afferent impulse, the stimulus was applied on the sub-petiole; the excitation had therefore to traverse the additional length of the sub-petiole, the conductivity of which is about ten times less than that of the petiole. The intensity of excitation is moreover considerably weakened during passage through the sub-petiole. This reduction in intensity, together with the added distance of sub-petiole to be traversed, explains the comparatively long time necessary for the transmission of the afferent impulse.

I have succeeded in shortening the transmission-time of the afferent impulse, by eliminating the intervening length of sub-petiole as in the following experiments.

Experiment 93.—The excitability of the pulvinus of the sub-petiole of Mimosa is considerably increased under favourable conditions of season and warmth in summer, when it readily responds to excitation by a lateral movement, as also by desiccation (*cf.* p. 69). By the application of glycerine the excitability of the secondary pulvini of the first and second sub-petioles was considerably increased. In effecting stimulation, the proximal electrode of the secondary coil was placed on the secondary pulvinus of the first sub-petiole, the second electrode on the middle of it. The impulse now travelled through the length of the petiole, and caused the fall of the leaf; the reflected impulse travelled outwards and caused the lateral movement of the secondary pulvinus of the second sub-petiole. The loss of time due to transmission through the short length of the sub-petiole was thus eliminated. The results of three typical experiments are given below. The afferent transmission-time is represented by t_1, that of the efferent transmission by t_2; V_a is the velocity of the afferent, and V_e that of the efferent impulse.

I. Length of petiole = 21 mm.
$t_1 = 1\cdot5$ sec.; V_a = 14 mm. per second.
$t_2 = 0\cdot5$ sec.; V_e = 42 mm. per second.

II. Length of petiole = 22 mm.
$t_1 = 2$ sec.; V_a = 11 mm. per second.
$t_2 = 0\cdot5$ sec.; V_e = 44 mm. per second.

III. Length of petiole = 51 mm.
$t_1 = 3$ sec.; V_a = 17 mm. per second.
$t_2 = 1$ sec.; V_e = 51 mm. per second.

The results described above show: first, that when the intermediate length of the sub-petiole is eliminated, the transmission-time of the afferent impulse is reduced to a value corresponding to that of the velocity of impulse in the petiole; and secondly, that the efferent impulse travels at least three times more quickly than the afferent. In reality it is even quicker; it was however impossible, by mere ocular observation, to carry quantitative measurement to a very high degree of accuracy, requiring as it does measurement of fractions of a second.

This very promising method of time-measurement by the indications of mechanical response of the secondary petiole is unfortunately not suitable for prolonged experiments; for long application of glycerine is apt to produce loss of excitability in the secondary pulvinus.

Summary

The definite nerve-connection between the nerve-end in each quadrant of the pulvinus at the centre and the corresponding sub-petiole at the periphery is demonstrated by localised central stimulation.

Feeble peripheral stimulation gives rise to the response characteristic of the corresponding quadrant of the pulvinus: the excitation remains localised. But when the intensity of peripheral stimulation is adequate, the in-going or afferent impulse reaching the pulvinus becomes reflected along a new path, as an efferent impulse. A Reflex Arc is formed at the centre.

The velocity of the efferent impulse is, in all cases, greater than that of the afferent impulse.

Experiments are described which show that the Reflex Arc is localised at the central end. Stimulation of sub-petiole (1) or sub-petiole (4) gives rise to a single reflected impulse; stimulation of sub-petiole (2) or sub-petiole (3) gives rise in each case to two reflected impulses. These characteristic effects are explained by the mutual relation of the several nerve-ends in the pulvinus.

CHAPTER XVI

SEPARATE CONDUCTING NERVES FOR SENSORY AND MOTOR IMPULSES, AND THE 'LOST TIME' IN REFLEX

In an animal nerve conduction takes place in both directions; hence it might be inferred that the same nerve-fibre may conduct sensory impulse from the periphery to the centre and motor impulse from the centre to the muscle. It is, however, held that this is not the case, but that in a single nerve-trunk there are two kinds of fibres isolated from each other, one of which subserves the afferent and the other the efferent function.

In the petiole of Mimosa also, both the sensory and motor impulses are apparently conducted by the same nerve. For stimulation of sub-petiole (1) gives rise to an afferent or sensory impulse along nerve (1) which, after reflection at the centre, becomes an efferent or motor impulse along nerve (2). Further, the sub-petioles (1) and (2) are interchangeable; external stimulus may be applied to sub-petiole (2) instead of to (1), when nerve (2) carries the sensory while nerve (1) conducts the motor impulse. It would thus appear, at first sight, that while in the animal great specialisation has been effected in separating the sensory and motor elements in the same nerve-trunk, no such differentiation had been reached in the nerve of the plant.

A detailed consideration of the results, however, brings out certain significant differences between the afferent and efferent impulses, which suggest that they may be transmitted by distinct channels. *By afferent or sensory impulse is meant the particular one generated by an external stimulus, which impulse is converted into efferent or motor in the reflex arc at the centre. The efferent impulse may therefore be distinguished as the reflected impulse.* Now there is a marked difference in the velocity of the two impulses by which they may be discriminated from one another: the efferent impulse is six to seven times quicker than the afferent. To take a concrete example: sub-petiole (1) is stimulated; the transmission-time to the centre along nerve (1) is 18 seconds; but when this impulse becomes efferent after reflection, it travels along nerve (2) in the course of 3 seconds. Sub-petiole (2) is now stimulated; the afferent impulse in nerve (2) takes 23 seconds to travel to the centre, in place of the 3 seconds taken by the efferent impulse to travel through the same distance. *Hence there may be two conducting elements in the same nerve-strand, one of which conducts the efferent or motor impulse and the other the afferent or sensory impulse, the velocity of the former being about six times greater than that of the latter.*

The existence of two distinct conducting nerves in the same vascular bundle is strongly supported by the following evidence:

(1) Microscopical examination (figs. 9, 11, 12) shows two separate phloems, one external, the other internal, to the xylem.

(2) Investigation with the Electric Probe demonstrated two distinct nervous impulses, one conducted by the outer and the other by the inner phloem (fig. 50).

The proof of the existence of separate conducting elements would be complete if it could be shown that in the same vascular bundle the external phloem conducted the slower and the internal phloem the quicker impulse. That this is the case is demonstrated by the following experiments.

Experiment 94. I stimulated in succession the external and the internal phloem of nerve (3) situated in the upper quadrant of the pulvinus; nerve (3), it will be remembered, terminates in sub-petiole (3) at the periphery. The transmitted impulse generated at the centre was, as usual, manifested by the closure of the first pair of sensitive leaflets. On the application of a superficial scratch to stimulate the upper phloem, the transmission-time to the third sub-petiole was found to be 22 seconds. On recovery of the leaflets, a second stimulation was effected by thrusting the pin deeper in, so as to reach the inner phloem; the velocity was now found to be greatly enhanced, the transmission-time being reduced to 3 seconds. *The inner phloem is therefore the conducting tissue for the more rapid impulse.* I carried out ten experiments with different specimens, numbers I. to X., and obtained very uniform results, as shown in the following table:

TABLE X.—THE DIFFERENT RATES OF TRANSMISSION BY THE OUTER AND THE INNER PHLOEM.

Transmission-time	I.	II.	III.	IV.	V.	VI.	VII.	VIII.	IX.	X.	Mean
Stimulation of outer phloem	sec. 22	sec. 20	sec. 28	sec. 18	sec. 26	sec. 19	sec. 18	sec. 30	sec. 20	sec. 24	sec. 22
Stimulation of inner phloem	sec. 3	sec. 4	sec. 4	sec. 3	sec. 4	sec. 4	sec. 3	sec. 4	sec. 3	sec. 4	sec. 3·6

It is very remarkable that the average times of the afferent and efferent impulses due to central stimulation given in the above table should so closely agree with those obtained by peripheral stimulation (Table IX.). The results of numerous experiments carried out by different methods show that the velocity in the inner phloem is about six to seven times greater than in the outer. The results of different lines of investigation all tend to indicate that the afferent or sensory impulse is conducted by the external, and the efferent or motor impulse by the internal phloem.

In the diagram (fig. 79, a) the internal phloem is shaded dark to distinguish it from the external phloem in light outline. Stimulation of the external phloem by scratch-stimulus causes a slow outgoing impulse Se, inducing response of the pulvinules of the sensitive leaflets on the sub-petiole. A deeper prick stimulates not only the external, but also the internal phloem. The quicker impulse reaches the pulvinule and causes an earlier response. The thick line

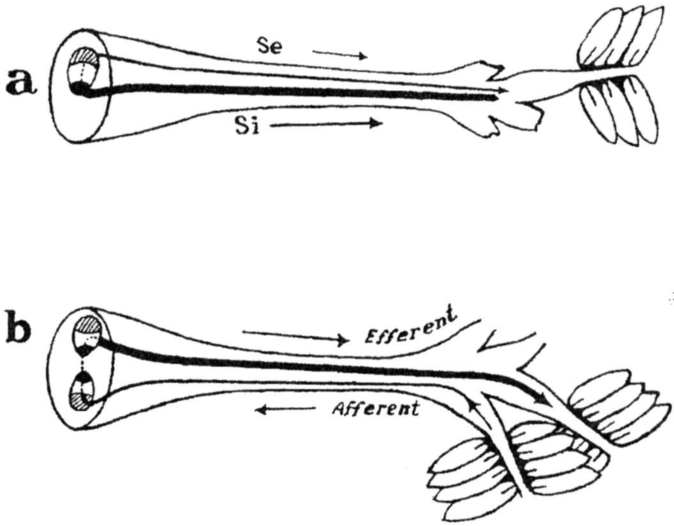

FIG. 79. Diagrammatic representation of transmission of impulses initiated by stimulation of external and internal phloem.

(a) Stimulation of outer phloem (shaded light) by superficial prick causes a slow outgoing impulse Se; deeper prick stimulates inner phloem (shaded black) and gives rise to a quicker impulse Si; the outer phloem-strand is represented by a thin line and the inner strand by a thick line; the lengths of the arrows represent the different velocities.

(b) Slow afferent impulse on stimulation of sub-petiole (1) is conducted along external phloem of nerve-strand (1); it passes into the inner phloem and then crosses over to inner phloem of nerve-strand (2). Quickened efferent impulse transmitted along inner phloem causes response in sub-petiole (2).

represents the inner nerve, and the quicker rate of conduction through it is represented by a longer arrow.

From the results of experiments already given, it will be seen that the descriptive terms, centripetal or afferent, and centrifugal or efferent, are to be regarded as relative and not absolute. For the same stimulus which applied at the periphery gives rise to a centripetal impulse, causes a centrifugal impulse when applied at the centre. The real difference lies in the conducting elements by which the two different impulses are propagated. The external phloem, which is the conductor for the slower impulse, is the 'sensory' nerve; whereas the inner phloem, which conducts the more intense and quicker impulse, is the 'motor' nerve.

The localisation of the external sensory and the internal motor nerves, affords a clearer insight into the transformation of the afferent impulse from the periphery into the efferent impulse at the centre. The inner phloems of the vascular bundles point inwards and are therefore in closest contiguity in the pulvinus. Fixing attention on the first sub-petiole, an external stimulus, under natural conditions, excites the outer conducting sensory nerve: this gives rise to a relatively slow afferent impulse in the external phloem which, reaching the pulvinus, passes inwards and thence to the inner phloem of nerve (2) which lies nearest, as shown in figure 79, b. This transference of excitation into the inner conductor involves an enhancement of speed characteristic of the efferent or motor impulse. The transformation of the afferent into efferent impulse in the reflex arc at the centre does not merely connote a reversal in the direction of propagation; for there is a marked disproportion between the afferent and the efferent impulses: it would appear as if a discharge of energy took place at the centre, by which the latter became far more intense than the feeble afferent impulse that provoked it.

I have hitherto described the effects of a moderate stimulus, under which the irradiation of excitation takes place at the centre from one nerve-end to the next. In the animal, the resistance to the passage of the impulse is overcome under a stronger intensity or longer duration of stimulus, in consequence of which there ensues a more widespread response. I now proceed to show that effects parallel to these are obtained in the reflex of Mimosa.

Successive Reflexes at Centre under Stronger Stimulation

In order to avoid unnecessary repetition, I describe in detail two typical series of experiments on the effect (1) of increase of intensity, and (2) of duration of stimulus.

Experiment 95. The experiments in this series were carried out with the same specimen, the intensity being gradually increased from minimum to maximum, the duration of application being kept constant, namely, one second.

(i) Tetanising induction-shock of ·5 unit was applied to sub-petiole (1). The reflected impulse only reached sub-petiole (2). There was thus a single reflex.

(ii) The intensity of the induction-current was now increased to 1 unit. The reflected impulse at the centre caused response not only of sub-petiole (2) but of sub-petiole (3) as well. There were thus two successive reflexes at the centre, under a moderate increase of intensity of the stimulus.

(iii) The stimulus was next increased to 3 units; three successive reflexes now occurred at the centre, in consequence of which the sub-petioles (2), (3) and (4) responded one after another. The reflected impulse reached the sub-petiole (2) 3 seconds after the fall of the leaf; the sub-petiole (3) responded 4 seconds after the response of (2); while the response of (4) was manifested 6 seconds after that of (3). There was always this sequence of responses in the order 2—3—4.

Similar effects were produced under an increased intensity of stimulus, when the fourth sub-petiole, to the extreme right, was subjected to stimulation. The sequence of response of the sub-petioles was now in the reverse order, namely, 3—2—1.

I will now describe the effect of increasing the duration of stimulation, the intensity remaining constant. The experiment was varied by the application of the stimulus to sub-petiole (4) on the extreme right.

Experiment 96.

(i) The duration of stimulation was at first 1 second. The impulse after reflection at the centre only reached sub-petiole (3). The reflex was therefore single.

(ii) The duration of stimulation was increased to 2 seconds; response occurred in both the sub-petioles (3) and (2), that of (2) being the later. There were therefore two successive reflexes at the centre.

(iii) The duration of stimulation was next increased to 4 seconds. This gave rise to successive responses of sub-petioles (3), (2) and (1): sub-petiole (2) responded 6 seconds after that of (3), and sub-petiole (1) 13 seconds after the response of (2).

It is thus apparent that increasing intensity or duration of stimulation elicits in Mimosa a corresponding outspread of response.

The 'Lost Time' in Reflex

The total delay between the application of stimulus to the first sub-petiole and the response of the second is made up of (*a*) the time of transmission of the afferent impulse from the periphery to the centre; (*b*) *the actual 'lost time' in crossing from one nerve to another at the centre,* in the first reflex; (*c*) the time of efferent transmission from the centre to the periphery; and (*d*) the latent period of the first sensitive leaflet which serves as the motile indicator. The actual lost time (*b*) at the first reflex is difficult to determine, since the values of (*a*), (*c*) and (*d*) are not known with sufficient exactitude.

As the length of the petiole through which the reflected impulse has to traverse is practically the same in all cases, the time-interval between the responses of sub-petioles (3) and (2) gives a fairly accurate measure of the time lost in the second reflex; similarly the time-interval between the responses of sub-petioles (4) and (3) gives the lost time in the third reflex.

TABLE XI.—THE LOST TIME IN REFLEX

Specimen	Lost time in 2nd reflex	Lost time in 3rd reflex	Total lost time T in 2nd and 3rd reflexes
1	4 seconds	6 seconds	10 seconds
2	4 seconds	8 seconds	12 seconds
3	6 seconds	13 seconds	19 seconds

The table above gives the lost time in successive reflexes in three different specimens. The lost time in the second reflex, under stimulation of moderate intensity, is in most cases 4 seconds; it is however longer in the third reflex, varying from 6 to 13 seconds. This difference is probably due to the fact that a decrement occurs in overcoming the block at the third reflex. This explanation derives some support from the success which often attended my attempt to reduce the lost time to the normal by previous treatment. This treatment consisted in applying a preliminary stimulus to the fourth sub-petiole, the after-effect of which resulted in the partial removal of the block that existed between the third and the fourth nerve-endings at the centre. The stimulation of the first sub-petiole now gave approximately similar values of lost time in the second and third reflexes. In ordinary circumstances, the total lost time T in the second and third reflexes is 10 seconds; owing to decrement it is often as long as 21 seconds.

I may now briefly recapitulate the effect of increased intensity of stimulus on response. *Minimal* stimulus applied to a sub-petiole reaches only the particular quadrant of the pulvinus with which the sub-petiole is in nervous communication; it does not proceed any further. The response induced is that characteristic of the particular quadrant reached by the stimulus. When the stimulus is increased from *minimal* to *moderate* intensity, it does not remain localised in the one quadrant, but becomes diffused, causing the fall of the leaf by stimulating to contraction the more excitable lower half of the pulvinus. A reflex arc is formed by the crossing over of the excitatory impulse from one nerve-end to the next; the reflected efferent impulse travels along a new path and causes response of the leaflets of the next sub-petiole. A still stronger stimulus becomes more widely irradiated, all the sub-petioles exhibiting response in serial succession. These characteristic features of the nervous reaction of Mimosa under increasing intensity of stimulus are extraordinarily similar to those of the nervous reaction in the animal, as summarised in the following quotation:

'A small stimulus will discharge only along the few neurons where the resistance is lowest. Increase of the stimulus, either by increase of its strength or by summation of weak stimuli, will enable the impulse to spread along more neurons and therefore will elicit a more widespread response. Only when the "blocks" are entirely removed by the administration of strychnine, or when the stimuli are abnormally powerful and long continued, will the impulse spread to all regions of the central nervous system, so that the response becomes general and inco-ordinate, instead of local and adapted to the stimulus.' [1]

The question now arises, whether the characteristic action of strychnine on the reflex of the animal can also be observed in the reflex of Mimosa? In my previous works I have demonstrated the very marked similarity of the effect of drugs on plant and animal tissues. Yet it at first appeared highly improbable that a vegetable alkaloid like strychnine should exert a similar action on plant and on animal nerves. The results of some of the successful experiments on the action of strychnine on plant-nerve were as astonishing as they were unexpected.

There are, however, certain inherent difficulties in working with the plant which render experimental success a matter of some uncertainty. Such are the fatigue

[1] Starling—*Principles of Human Physiology*, 1920, p. 305.

induced by prolonged experimentation, and the absorption of the drug in excess of the proper dose. Both these factors tend to cause a prolongation of the lost time in reflex, as was ascertained in a large number of experiments. But, under favourable circumstances, the lost time was found to be reduced to zero, which cannot be accounted for by any other supposition than that of the abolition of the block under the characteristic action of strychnine. The results of some twenty successful experiments were too consistent to be ignored.

EFFECT OF STRYCHNINE

I will first describe the test by which the effect of strychnine on the reflex is to be proved. It has been shown that the total lost time T in the second and the third reflexes varied from 10 to 21 seconds. If strychnine were to cause a more or less complete removal of the block, the fact would be demonstrated by a reduction or abolition of the lost time.

The greatest difficulty encountered was in the administration of the proper dose of strychnine to the plant, any excess proving to be highly toxic. In the animal, a small and definite dose can be easily injected, taking the precaution that the dose is not in excess. There is no such facility for the plant, the only practicable method being the application of a dilute solution of strychnine hydrochloride on the pulvinus, which after absorption would reach the central nerve-endings: but the absorption of a proper quantity or dose is a question more or less of chance. Absorption of too small a dose was found to produce *diminution*, but not complete abolition, of the lost time; whereas absorption of the proper quantity produced a *complete removal of the block*, the lost time being reduced to zero. Excessive absorption, on the other hand, caused an *increase* of lost time. A moment's consideration will show how variable are the factors which modify the rate of absorption: it depends on the absorptive power of the epidermis, the physiological activity and the turgor of the plant: it is also affected by the prevailing temperature and hygrometric conditions.

There is a limitation in point of time for the observation of the effect of the duration of an application of strychnine. Mimosa requires a period of some 15 minutes for full recovery of its conductivity and excitability after each stimulation. Hence if the effect of 5 minutes' application of strychnine be first observed, the next observation can only be taken at the twentieth minute, and the third at the thirty-fifth minute. If any of these periods happen to coincide with the critical period of absorption of the proper dose, only then would a complete abolition of the block be indicated by the simultaneous response of all the subpetioles. In spite of all these difficulties, a fair number of experiments were successful, which showed that application of the proper dose of strychnine produces complete abolition of the block. I give below detailed account of some of the typical experiments carried out with exceptionally vigorous specimens.

Experiment 97.—The total lost time T in the second and third reflexes was found to be 12 seconds; after application of dilute solution of strychnine, 1 part in 10,000, for 5 minutes, the block was partially removed, the lost time being reduced from 12 to 5 seconds. The next observation at the twentieth minute showed *complete removal of the block*, the responses of sub-petioles (2), (3) and (4) being simultaneous. Continued application produced a toxic effect; for at the thirty-fifth minute the lost time increased from zero to 9 seconds: longer application abolished all response. The curve obtained from the data given below is given in fig. 80, *a*.

	Lost time T.
Normal	12 seconds.
After 5 minutes' application of strychnine .	5 ,,
,, 20 ,, ,, ,, .	0 ,,
,, 35 ,, ,, ,, .	9 ,,

In this case the period of critical absorption was 20 minutes: but under conditions favourable to more rapid absorption, an application of 20 minutes would pro-

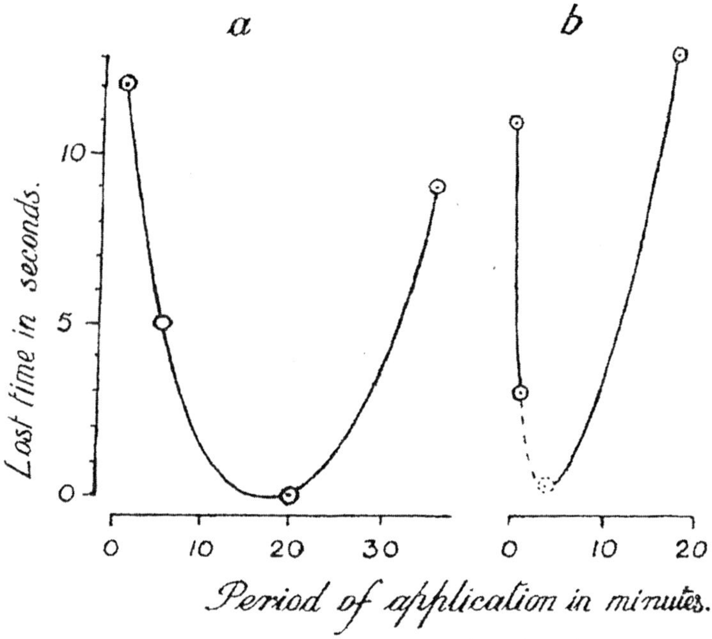

FIG. 80. Curve of relation between duration of application of strychnine and the lost time.

The critical period in curve (*a*) is about 20 minutes, in (*b*) it is shortened to 2 minutes. (See text.) Duration of application in minutes, and lost time in seconds.

duce a toxic effect and prolong the lost time. This accounts for the result obtained a few days later (May), when the temperature was higher than usual, that the application of strychnine even for 5 minutes prolonged the lost time. I therefore took an observation with a fresh specimen a minute after the application; the lost time was now found to be reduced from the normal 13 seconds to 2 seconds only; an observation taken 15 minutes afterwards (duration of application 16 minutes) showed that the lost time after reaching a minimum had become prolonged to 9 seconds. The critical period in this case was probably 2 minutes (see fig. 80, *b*).

I give two other examples of experiments carried out in autumn (September) when the rate of absorption was comparatively slow. I therefore employed a stronger solution, one part in a thousand, and observed the effect of increased duration of the application of strychnine.

Experiment 98.—The normal lost time was found to be 15 seconds. The application of strychnine for 2 minutes caused complete abolition of the block, the responses of sub-petioles (2), (3) and (4) being simultaneous. After an additional interval of 15 minutes, retardation occurred, the lost time being prolonged to 8 seconds. Absorption of strychnine for a further period of 15 minutes abolished all response.

Experiment 99.—The variation was introduced of stimulating the fourth instead of the first sub-petiole; the sequence of response was therefore (3), (2) and (1). The normal lost time was 12 seconds. After application of strychnine for 2 minutes, the block was completely removed as shown by the simultaneous responses of sub-petioles (3), (2) and (1). Prolonged application for a further period of 15 minutes caused a retardation, the lost time being now lengthened to 7 seconds.

The results given above indicate that under the proper dose of strychnine, the block becomes completely abolished, as observed in the reflex of the animal.

The nervous mechanism of Mimosa has been shown to be far more complex than had hitherto been suspected. The question now arises: What advantage in the economy of the plant is secured by such a highly elaborated nervous mechanism? One advantage, as already pointed out, is the co-ordinated reflex by which the leaf-surface is adjusted perpendicularly to the incident light so as to ensure the absorption of the optimum amount of radiant energy for photosynthesis. Another important function of the nervous mechanism is the transmission of a rapid message to the motile organ for a quick reaction to avoid a threatened danger. It has been suggested that the sudden fall of the leaf serves the purpose of scaring away grazing cattle. But the cow is not sufficiently intelligent to take notice of the movement of the leaf or to be frightened by it. The reflex movements may, however, subserve the protection of the plant in a different way. The long procumbent stem of *Mimosa pudica,* which covers large patches of ground in the tropics, bears numerous leaves. When one of the sub-petioles bearing leaflets is trampled on or bitten, the excitatory impulse is transmitted throughout the length of the plant. Confining our attention to a particular leaf, the afferent or sensory impulse causes a fall of the leaf which presses itself against the ground. The successive reflexes at the centre give rise to motor impulses which make the four sub-petioles approach each other laterally, and also cause closure of all the leaflets. Nothing could be more striking than the rapid change by which a patch of vivid green becomes transformed into thin lines of dull grey unnoticed against the dark ground. It is probable that this invisibility may serve as a means of protection.

Summary

When a stimulus acting on one of the sub-petioles is increased above the minimal intensity, the afferent impulse becomes reflected and transformed at the reflex-centre in the pulvinus into an efferent impulse which travels along a new path. This connotes not merely a reversal in direction, but also a considerable discharge of energy ; for there is a marked disproportion between the afferent and efferent impulses, the latter being the more intense as evidenced by its higher velocity of transmission.

Two distinct conducting elements have been localised

in the nerve-strand of Mimosa, of which the outer conducts the sensory and the inner the motor impulse.

The impulse becomes more widespread under stronger intensity of stimulus.

The resistance to the passage of impulse at the centre is completely removed by the absorption of the proper dose of strychnine. The lost time is reduced to zero, as is shown by the simultaneous response in all the sub-petioles.

CHAPTER XVII

GENERAL REVIEW

THE irritability of an unicellular organism is exhibited by its contractility, rhythmicity and conductivity, which are different aspects of a fundamental protoplasmic reaction. These characteristics of the individual cell persist even when cell-complexes and tissue-systems are formed in multicellular organisms. In the physiological division of labour, each particular tissue becomes specialised to discharge a given function in a more efficient manner. Thus whilst all tissues are more or less contractile, high sensitiveness and power of rapid contraction have been attained by the pulvinus of Mimosa, which responds to stimulus of an intensity even below the threshold of human perception. I have shown elsewhere that rhythmic or pulsatory activity is exhibited very widely by different tissues, and that a continuity exists between multiple and autonomous response. The autonomous activity is conspicuously manifested by the pulsating leaflet of *Desmodium gyrans*; but it requires appliances of high sensitivity to detect it in growth and in the activity which maintains the ascent of sap. There is similar continuity in the conductivity of the different tissues; the indifferent tissues are feeble conductors, while the tubular cells in the phloem-strands are most effective in conducting, even a feeble excitation, to a distance.

The intercommunication and interaction of distant organs of the plant are secured (1) by slow translocation of chemical stimulants by the movement of sap, and (2) by rapid transmission of protoplasmic excitation corresponding to the nervous impulse in the animal. Great confusion has arisen in the study of the conduction of excitation in plants from want of discrimination between the two types of action at a distance, physical and physiological. Still more serious has been the error introduced into the investigation by the application of too violent stimuli. The intense excitation thus produced remains no longer confined to the conducting tissue of the plant or the animal, but becomes widely diffused (pp. 12, 41, 195).

Theory of Hydro-Mechanical Conduction

This theory, advocated by Pfeffer and Haberlandt, is based upon the observation of the escape of a drop of water from the petiole of *Mimosa pudica* after a deep cut. The sudden variation of hydrostatic pressure thus produced is supposed to be hydro-mechanically transmitted to a distance so as to inflict a blow on the excitable pulvinus, with resulting fall of the leaf. The inadequacy of this theory is evident from the fact that stimulation of the plant, and the transmission of excitation to a distance, can be effected by scratch-stimulus, by superficial friction, or by an electric shock of feeble intensity. Transmission of excitation thus takes place without any wound and resulting escape of water which might have produced a hydro-mechanical disturbance.

Theory of Transpiration-Current Conduction

This is based on the supposition that stimulation is produced by some chemical irritant excreted in consequence of irritation of the wood, and that the translocation of the irritant to the distant leaf is effected by the movement of sap, the rate of which is assumed to be the same as that of the transmission of excitation. The baselessness of the theory is shown:

(1) By the initiation and transmission of excitation without in any way irritating the wood (p. 14).

(2) By the simultaneous transmission of excitation upwards and downwards, with and against the direction of the transpiration-current; also by the conversion of the ascending into a descending impulse after crossing over at the apex of the stem (p. 4).

(3) By the results of experiments in which a stimulant was applied at the extreme tip of the leaf, or of a flower-stalk; the excitatory impulse generated travelled to a considerable distance downwards against the direction of the normal ascent of sap; subsequent examination showed that the stimulant had not been transported, but had remained localised at the point of application (p. 18).

The theory of conduction by the transpiration-current is thus completely discredited, conduction of excitation being in no way connected with the movement of sap.

The transmission of excitation is neither hydromechanical nor due to the movement of sap. The following crucial experiment proves that conduction in the plant is a propagation of protoplasmic excitation like the nervous conduction in the animal.

Polar Excitatory Effect of Constant Current

The effect of an electric current is discriminative in its action. In the conducting nerve of the animal, a feeble current excites only on make at the kathode; under a stronger current, excitation occurs on make at the kathode, and on break at the anode. My experiments on the polar action of an electric current prove that the reactions in the plant are identical with those in the animal. Excitation, under a feeble current, is produced only at kathode-make, and the excitatory impulse thus generated is transmitted to a distance. Under a stronger current, excitation is induced both at kathode-make and at anode-break (p. 25). These results offer conclusive proof of conduction in plants being a physiological propagation of protoplasmic excitation as in the animal.

Automatic Determination of Velocity of Transmission

For investigating the velocity of transmission of excitation in plants, it is essential to devise a method of determination of as high a degree of accuracy as that employed for the study of the nervous impulse in the animal. This has been secured by my Resonant Recorder, in which errors arising from frictional resistance in recording have been completely removed. The dotted record permits estimation of time-intervals as short as 0·001 second.

Latent period.—The average value of the latent period of the pulvinus of Mimosa is 0·1 second. It is shortened, within limits, with rise of temperature; it is prolonged under fatigue.

Velocity of transmission.—The velocity in thick petioles of Mimosa, in summer, varies from 30 to 55 mm. per second. In thin petioles it is as high as 400 mm. per second (p. 64). The velocity of impulse in Mimosa is not as great as in the higher, and not so low as in the lower animals. The velocity of conduction in the sub-petiole and in the main stem of Mimosa is considerably lower than in the petiole.

Influence of Tonic Condition on Conduction

The velocity is modified by the physiological condition of the plant. In winter it is very much lower than in summer.

If the plant be kept in unfavourable condition, its physiological tone falls *below par*, when the conducting power becomes depressed or arrested. Strong stimulation is found to renew or enhance the power of conduction. The effect of intense stimulation on conduction in normal tissues is exactly the opposite, *i.e.* a depression or fatigue of conduction (p. 66).

Physiological Block of Conduction

Arrest by electrotonic block.—The nervous impulse in an animal nerve is arrested by the interposition of an electrotonic block, the conduction being restored on the cessation of the electric current. Similarly, transmission of impulse in the plant can be repeatedly arrested or restored by the alternate application and removal of the electrotonic block (pp. 29, 75).

Block of conduction by local cooling.—The animal nerve, when excessively cooled, loses its power of conduction. A parallel effect is observed in the nervous conduction in the petiole of Mimosa. Local application of ice paralyses the conducting power, which persists for a time even after the return to a normal temperature. The paralysis is quickly removed by electric stimulation of the benumbed tissue (p. 74).

Abolition of conduction by the local action of poison.—The application of poisonous solutions on a narrow zone in the petiole of Mimosa induces a depression of the power of conduction, which culminates in a permanent abolition. The time required for the abolition depends on the virulence of the poison employed, being quicker under potassium cyanide than under copper sulphate solution (p. 77).

These results conclusively prove that the transmission of excitatory impulse in plants is essentially similar to that of the nervous impulse in animals.

Anatomical Characteristics of the Conducting Tissue

The phloem in the petiole is now admitted by some observers to be the tissue that conducts excitation. On the supposition that protoplasmic continuity is essential for conduction, the sieve-tubes, possessing numerous perforations across the transverse septa traversed by threads of protoplasm, were regarded as the special conductors of excitation. This supposition is, however, erroneous for two reasons: (1) the sieve-tubes discharge a different function; and (2) the conducting phloem consists mainly of tubular cells with imperforate septa. Protoplasmic continuity is by no means essential to the conduction of excitation, for in the animal there is no such continuity across a nerve-junction, where neurone joins neurone, the separating membrane being known as a synapse. Similar synapsoidal membranes are shown to exist between contiguous conducting cells in the plant.

The anatomical structure of the conducting phloem in the stem is in every way similar to that in the petiole; both contain the tubular cells which are effective in conduction.

Distribution of the Conducting Tissue

Main conducting strands in the stem.—There are two opposite main phloem-strands in the stem which converge and meet at the apex (p. 40). This particular distribution explains why, under unilateral stimulation of the stem, the ascending impulse crosses over at the apex and becomes reversed at the opposite side (p. 5).

Nerve-connection between stem and leaves.—The two main phloem-strands in the stem give off lateral branches to the leaves, thereby assuring conducting continuity between the stem and the leaves. The impulse initiated by the stimulation of the stem is propagated in a centrifugal direction to the leaves, while the excitation initiated by the stimulation of the leaf travels centripetally to the stem and is there conducted up and down, causing the fall of other leaves borne on the stem.

Distribution of conducting strands.—It is difficult to trace the course of the nerve-strands in the plant; but on staining with hæmatoxylin, the conducting phloem becomes stained deep violet against the surrounding tissue. The phloem in each bundle is not single but double, the xylem being interposed between the two. The bundle is in fact bicollateral (p. 37).

SYNAPSE AND SYNAPSOID

The imperforate septa between the tubular conducting cells act as synapsoidal membranes. The effect of the interposition of synapses in the path of nervous impulse in the animal is characterised as follows: (1) Conduction is irreciprocal or unequal in two directions: (2) Under moderate stimulation, the resistance or block to the passage of impulse is diminished, a phenomenon known as 'Bahnung' or Facilitation: (3) Under excessive stimulation, an increase in the block occurs at each synapse, producing a fatigue of conduction.

Irreciprocal conduction.—At the neuro-muscular junction of the animal, conduction is irreciprocal; there is no obstacle to prevent the impulse from the nerve passing into the muscle, but the excitation of the muscle does not pass backwards into the nerve. Similarly, at the junction of the conducting and the contractile tissues of Mimosa, conduction is irreciprocal. Excitation of the conducting nerve passes easily into the pulvinus, but the contraction of the pulvinus does not affect the enclosed nerve-endings (p. 45).

Continuity between irreciprocal and preferential conduction.—Under a minimal stimulus, conduction is irreciprocal in various sensitive plants, *i.e.* it takes place in one direction and not in the opposite. Under increasing intensity of stimulus the synapsoidal block becomes more and more effectively forced. Conduction is, however, quicker in the preferential direction (p. 48).

'Bahnung' or Facilitation occurs as an after-effect of moderate stimulation. A formerly ineffective stimulus now becomes fully effective. Stimulus thus canalises its own conducting path (p. 51).

Fatigue of conduction.—After excessive or long continued stimulation, conduction exhibits a depression or fatigue.

Electric Response to Transmitted Excitation

The excitatory reaction has two concomitant expressions: the mechanical response of motile organs, and the electric response of all tissues, motile or non-motile. The fact that galvanometric negativity is a concomitant of excitatory reaction is proved by simultaneous observation of mechanical and electrical response of Mimosa to transmitted excitation, when the two responses are seen to take place practically at the same time (p. 105). The electrical and mechanical responses are independent manifestations of the common excitatory process, for the response by galvanometric negativity takes place even when the leaf is restrained from physical movement. The fact that the excitatory process is independent of the process of contraction is proved by the persistent excitatory electric response of the pulvinus when rendered immotile by absorption of excess of water (p. 107). The transmission of excitation in ordinary plants is proved by the electric response, since it takes place even in non-motile tissues. The invisible nervous impulse is detected by the electric change of galvanometric negativity which accompanies it.

Localisation of the conducting nerve in the petiole of Mimosa by the Electric Probe.—On thrusting in the Probe perpendicularly to the diameter of the petiole by steps of 0·05 mm., an excitatory transmission inducing strong galvanometric negativity was found to occur only when one or other of the two phloems in each bundle, the external or the internal, was reached (p. 113). The two phloems are therefore the conductors of excitation. The existence of the two phloems in each bundle was also demonstrated by selective staining.

Determination of the velocity of transmission by diphasic electric response.—Two electric contacts were made with the conducting nerve in the petiole of Mimosa and record obtained with the Einthoven string-galvanometer. The excitatory impulse arriving at the proximal contact A induced electric negativity of that point; the later arrival of the impulse at the distal point B produced a reversal of the recorded curve of response. The time-interval between the two phases of response, and the distance between A and B, sufficed for the determination of the velocity, which was the same as that obtained by the mechanical method (p. 109).

Electric response to transmitted excitation in ordinary plants.—Examination of the midrib of leaves showed that the phloem contains tubular cells similar to those which conduct excitation in Mimosa. The transmitted excitation along the conducting phloem has been detected by electric response of galvanometric negativity. The various characteristics of conduction evidenced by mechanical and electric response of Mimosa have also been found in the electric response of ordinary plants. The conduction is irreciprocal under feeble, and preferential under stronger, stimulation (p. 102). 'Bahnung' or Facilitation of conduction is observed, the removal of block by successive stimulations being demonstrated by a staircase-increase of the electric response (p. 101).

Nervous conduction takes place not only in the 'sensitive,' but in all plants.

Positive and Negative Impulses under Indirect Stimulation

Two distinct impulses are generated under indirect stimulation, a positive followed by a negative. The velocity of the positive impulse is greater than that of the negative: but the reaction to the positive is relatively slow and feeble, whereas that due to the negative is abrupt and intense.

Effects of positive impulse.—The positive impulse gives rise to an enhancement of turgor, causing (1) expansion and erectile response of the motile leaf or leaflet; (2) acceleration of the rate of growth of a growing organ; (3) an electric response of galvanometric positivity.

Effects of negative impulse.—The negative impulse is excitatory, and gives rise to a diminution of turgor, manifested by (1) contraction and responsive fall of the motile leaf; (2) retardation of the rate of growth; (3) an electric response of galvanometric negativity.

The resultant response is determined (*a*) by the intensity of the stimulus; (*b*) by the distance of transmission; (*c*) by the conductivity of the tissue; and (*d*) by the tonic condition. The two responses of opposite sign, a feeble positive followed by a stronger negative, are exhibited by a semi-conducting tissue. On reducing the intervening distance, the predominant negative overtakes and masks the feeble positive (p. 85). When the distance is reduced to zero, stimulation becomes Direct, and the resultant response is one of contraction, of retardation of the rate of growth, and of galvanometric negativity. Under the condition of sub-tonicity, when conductivity is depressed, the excitatory impulse is arrested and the positive impulse is alone transmitted.

The following is the Law of Effects of Direct and Indirect Stimulation:

> (1) THE EFFECT OF ALL FORMS OF DIRECT STIMULATION IS A DIMINUTION OF TURGOR, CONTRACTION, DIMINISHED RATE OF GROWTH, A NEGATIVE MECHANICAL RESPONSE AND AN ELECTRIC RESPONSE OF GALVANOMETRIC NEGATIVITY.
>
> (2) THE EFFECT OF ALL INDIRECT STIMULATION OF FEEBLE OR MODERATE INTENSITY IS AN INCREASE OF TURGOR, EXPANSION, ACCELERATION OF RATE OF GROWTH, A POSITIVE MECHANICAL RESPONSE AND AN ELECTRIC RESPONSE OF GALVANOMETRIC POSITIVITY.

ACCESSION AND DEPLETION OF ENERGY UNDER THE ACTION OF STIMULUS

Stimulation does not always cause a depletion or run-down of energy D; under certain conditions it causes an accession and storage of energy A. The relative intensity of the two reactions is modified in a definite manner according to the tonic condition of the tissue; when the tonic level is *above par*, the D reaction is predominant (D > A); but when the tonic condition is *below par*, the accession of energy is the more pronounced (A > D), the tonic level being thereby raised to the normal. Under repeated stimulation the arrested power of conduction is restored by the gradual removal of the block offered by the synaptic membranes. The fundamental protoplasmic reaction under the action of stimulus shows parallel changes in the three manifestations of irritability. In a condition of sub-tonicity the motor-excitability, conductivity and rhythmicity are all depressed or arrested; the arrested activities become revived under stimulation. The effect of stimulation on tissues in an optimum condition is the opposite of that on sub-tonic tissues, depression or fatigue being induced in the various manifestations of irritability.

RESPONSE OF THE ISOLATED PLANT-NERVE

The nerve of the Fern can be isolated without injury; it is a vascular strand which includes conducting phloem consisting of tubular cells. The characteristics of conduction in the plant-nerve are in every way similar to those in animal nerve, both in normal and modified conditions. In the normal nerve, the negative impulse masks the positive, and the resultant response to indirect stimulation is negative. Conductivity is depressed or temporarily abolished under a condition of sub-tonicity; of the two impulses the excitatory negative is now arrested, and the positive alone is transmitted. This explains the abnormal positive response of animal and plant nerves to indirect stimulation. The block to the passage of excitatory impulse is gradually removed under successive stimulations, and the response becomes diphasic, positive followed by negative. The conductivity is fully restored after a period of tetanisation or continuous stimulation, and the abnormal positive becomes converted into the normal negative response (p. 128). Since the results are identical in plant and animal nerves, the physiological mechanism must be the same in both.

Directive Action of Electric Current on Conduction

The identical nature of the physiological mechanism of nervous conduction in plant and animal is further demonstrated by similar modifications of conduction under the directive action of a constant electric current. A current is *homodromous* when its direction is the same as that of the propagated nervous impulse; it is *heterodromous* when opposed to the direction of propagation.

Effect of heterodromous current.—The velocity of transmission is enhanced when against the direction of the current. The intensity of a transmitted excitation is also increased. Ineffectively transmitted excitation becomes effectively transmitted (p. 141).

Effect of homodromous current.—The velocity of transmission is depressed. An effectively transmitted excitation becomes arrested during the passage of the current (p. 141).

Immediate after-effect.—A 'rebound' takes place immediately after the stoppage of the current. The result is a transient variation of conductivity, whose sign is opposite to that induced during the passage of the current. The transient after-effect of a heterodromous current is a depression of the normal conductivity; that of a homodromous current, an enhancement of conductivity above the normal (p. 142).

There is a critical intensity of current above which the above effects undergo a reversal.

The following are the Laws of Conductivity Variation in plant and animal nerves, under electric current of feeble intensity:

(1) HETERODROMOUS CURRENT ENHANCES, AND HOMODROMOUS CURRENT DEPRESSES OR ARRESTS THE CONDUCTION OF EXCITATION.

(2) THE AFTER-EFFECT OF THE CURRENT IS A TRANSIENT VARIATION OF CONDUCTIVITY, THE SIGN OF WHICH IS OPPOSITE TO THAT INDUCED DURING THE PASSAGE OF THE CURRENT.

COMPLEXITY OF THE MOTILE ORGAN

The four quadrants of the pulvinus of Mimosa act as four different effectors with characteristic responsive leaf-movements. Under direct stimulation, the upper quadrant responds by an up-, and the lower quadrant by a down-movement; the response of the left quadrant is by an anti-clockwise, and of the right quadrant by a clockwise torsion (p. 155).

The Law of Torsional Response IS THAT AN ANISOTROPIC ORGAN LATERALLY EXCITED BY STIMULUS, UNDERGOES TORSION SUCH THAT ITS LESS EXCITABLE SIDE IS MADE TO FACE THE STIMULUS.

NERVOUS CONTROL OF THE ATTITUDE OF THE LEAF

The surface of the leaf places itself at right angles to the direction of the incident light. In Mimosa, the motor organ is at some distance from the perceptive lamina; the attitude of the leaf is modified by separate nervous impulses transmitted from the four sub-petioles bearing the leaflets when stimulated by light. The definite nerve-connection between each of the sub-petioles and the corresponding quadrant of the pulvinus is proved independently (1) by peripheral and (2) by central stimulation.

Characteristic effects of peripheral stimulation.—Stimulation of sub-petiole (1), by moderate electric or photic stimulation induces an anti-clockwise torsion, similar to that produced by direct stimulation of quadrant (1). Sub-petiole (1) must therefore be in nervous connection with quadrant (1) of the pulvinus. Similarly, sub-petiole (2) is shown to be connected with quadrant (2), sub-petiole (3) with quadrant (3), and sub-petiole (4) with quadrant (4) (p. 168).

Effects of central stimulation.—Stimulation of the central nerve-end in the left quadrant (1) generates an excitatory impulse, which travelling outwards causes response only in the leaflets of sub-petiole (1). Local excitation of the nerve-end in quadrant (2) causes response in sub-petiole (2). Stimulation of nerve-ends in quadrants (3) and (4) elicits corresponding response in sub-petiole (3) and in sub-petiole (4) (p. 174).

Co-ordinated reflex in leaf-adjustment.—When the leaflets of the right sub-petiole (4) are acted on by vertical light, the distant pulvinus undergoes a right-handed or clockwise torsion, by which the leaflets are carried away from a position at right angles to the incident light. But when the two sub-petioles (1) and (4) are simultaneously exposed to light of the same intensity, the two resultant torsions balance each other. The lateral adjustment of the leaf as a whole is thus made by the two sub-petioles (1) and (4) situated externally. The balancing adjustment up or down, is effected in response to excitations transmitted by the two middle sub-petioles. Equilibrium is attained when the leaf-surface as a whole is perpendicular to the incident light. The dia-heliotropic attitude of the leaf is thus brought about by distinct nervous impulses, initiated at the perceptive region actuating different effectors at a distance (p. 171).

The Reflex Arc

A minimal stimulus applied to one of the sub-petioles gives rise to an in-going impulse causing the characteristic response of the corresponding quadrant of the pulvinus; the excitation remains localised without irradiation to the other nerve-ends. Under a slight increase of the intensity of stimulus, a different class of phenomenon makes its appearance; the in-going or afferent impulse, reaching the centre, now becomes reflected along a new path as an efferent impulse. Thus a reflex arc is formed at the centre (p. 175).

Relation between the afferent and the efferent impulses.—Stimulation of sub-petiole (1) or sub-petiole (4) gives rise to a single reflected impulse; stimulation of sub-petiole (2) or (3) gives rise in each case to two reflected impulses. These characteristic effects are explained by the mutual relation of the contiguous nerve-ends at the centre (p. 183).

External stimulus applied at the periphery generates an afferent or sensory impulse, which, after reflection at the centre, is transformed into an efferent or motor impulse. There is a marked difference in the velocity of the two impulses, by which they may be discriminated from each other, the motor impulse being about seven times as quick as the sensory (p. 182). The transformation of the sensory into the motor in the reflex arc connotes not merely a reversal in the direction of propagation, but also a great discharge of energy at the centre, by which the motor impulse becomes far more intense than the feeble sensory impulse that provoked it.

Separate conducting nerves in the same bundle.—Direct evidence is adduced in demonstration of the existence of two separate nerves for the motor and the sensory impulses, the characteristic of the motor nerve being that conduction takes place in it at a much quicker rate than in the sensory.

The fact that there are two conducting phloems, one external and the other internal, in the same bundle has been proved by differential staining; tubular cells have also been found in both (p. 36). The separate transmission of excitation by the two phloems has been independently demonstrated by investigations carried out with the Electric Probe (pp. 113).

Conduction of sensory impulse by the external, and of the motor impulse by the internal phloem.—This is proved by separate and successive stimulations of the external and of the internal phloem. Stimulation of the external phloem gives rise to a relatively slow impulse, which is characteristic of the sensory impulse; stimulation of the internal phloem gives rise, on the other hand, to an impulse which is seven times quicker, which is characteristic of the motor impulse (p. 189). The difference between the two impulses does not lie in the direction of propagation. For stimulation of the external phloem at the periphery and at the centre both give rise to a slowly transmitted impulse, though in the former case the direction of propagation is in-going, and in the latter, out-going. The real difference lies in the conducting nerves. Under natural conditions, the afferent impulse from the periphery is conducted to the centre by the external phloem; the excitation is then transferred to the inner phloem of the contiguous nerve and conducted outwards as the efferent impulse (p. 191).

SUCCESSIVE REFLEXES UNDER STRONGER STIMULUS AND THE 'LOST TIME' IN REFLEX

A more widespread response ensues on the application of a stronger stimulus. Stimulation of the left sub-petiole (1) then causes successive reflexes at the centre, the sub-petioles (2), (3) and (4) responding one after another in the sequence 2–3–4. Stimulation of the right sub-petiole (4) causes successive reflexes in the reverse order of 3–2–1.

The 'lost time' in reflex.—The time lost in the second reflex is normally about 4 seconds; on account of decrement, it is longer at the third reflex. The total lost time in the second and third reflex is from 10 to 19 seconds.

Abolition of 'block' under strychnine.—The block, causing 'lost time' in reflex in the animal nerve, is entirely removed after administration of strychnine. A similar effect is also observed in the reflex of Mimosa; the resistance to the passage of the impulse at the centre becomes completely removed by the absorption of the proper dose of strychnine. The 'lost time' is then reduced to zero, as is shown by the simultaneous response in all the sub-petioles.

In the light of the results summarised in this chapter, it can no longer be doubted that plants, at any rate vascular plants, possess a well-defined nervous system.

It has been demonstrated that excitation is conducted by the phloem of the vascular bundle, and that conduction in this tissue can be modified experimentally by the same means as is that in animal nerve. The conducted excitation may, therefore, be justly spoken of as nervous impulse and the conducting tissue as nerve.

It has been further shown that, as in the animal, it is possible to distinguish sensory or afferent and motor or efferent impulses, and to trace the transformation of the one into the other in a reflex arc. The observations involve the conception of some kind of nerve-centre. No structure corresponding to the nerve-ganglion of an animal has, indeed, been discovered in the pulvinus of *Mimosa pudica,* but it is not impossible that the physiological facts may one day receive histological verification.

INDEX

Published by

Facsimile Publisher
Ashok Vihar, Delhi-110052 (India)
Website: www.facsimilepublishers.com

Distributed by

Gyan Books Pvt. Ltd.
5, Ansari Road, Daryaganj, New Delhi-110002 (India)
E-mail: books@gyanbooks.com, Website: www.gyanbooks.com

9 789333 351096